THE RECORDING ENGINEER'S HANDBOOK

BOBBY OWSINSKI

FIFTH EDITION

The Recording Engineer's Handbook 5th edition
by Bobby Owsinski

Published by:
Bobby Owsinski Media Group
4109 West Burbank, Blvd.
Burbank, CA 91505

© Bobby Owsinski 2023
ISBN 13: 978-1-946837-19-6

ALL RIGHTS RESERVED. No part of this work covered by the copyright herein may be reproduced, transmitted, stored, or used in any form and by any means graphic, electronic or mechanical, including but not limited to photocopying, scanning, digitizing, taping, Web distribution, information networks or information storage and retrieval systems, except as permitted in Sections 107 or 108 of the 1976 Copyright Act, without the prior written permission of the publisher.

For permission to use text or information from this product, submit requests to requests@bobbyowsinski.com.

Please note that much of this publication is based on personal experience and anecdotal evidence. Although the author and publisher have made every reasonable attempt to achieve complete accuracy of the content in this Guide, they assume no responsibility for errors or omissions. Also, you should use this information as you see fit, and at your own risk. Your particular situation may not be exactly suited to the examples illustrated herein; in fact, it's likely that they won't be the same, and you should adjust your use of the information and recommendations accordingly.

Any trademarks, service marks, product names or named features are assumed to be the property of their respective owners, and are used only for reference. There is no implied endorsement if we use one of these terms.

Finally, nothing in this book is intended to replace common sense, legal, medical or other professional advice, and is meant to inform and entertain the reader.

To buy books in quantity for corporate use or incentives, email office@bobbyowsinski.com.

TABLE OF CONTENTS

Introduction 1

PART I

How Microphones Work 7

Common Microphones 27

Basic Recording Equipment 55

DAW Recording 73

Microphone Placement Fundamentals 87

Basic Stereo Techniques 99

Preparing The Drum Kit For Recording 107

Recording Drums 117

Miking Individual Instruments 149

Recording Basic Tracks 207

Recording Overdubs 221

Immersive Recording Techniques 231

PART II

Chuck Ainlay 245

Michael Bishop 251

J.J. Blair 257

Bruce Botnick 261

Wyn Davis 267

Eddie Kramer 273

Mack	277
Sylvia Massy	281
Dennis Moody	287
Barry Rudolph	291
Al Schmitt	297
Glossary	303
About Bobby Owsinski	315

INTRODUCTION

When the first edition of this book was written back in 2002, the recording world was a way different place. There were still quite a few real commercial studios available to record in, the old studio structure of experienced engineer teaching the assistant was still in place, and you still wanted to be signed to a record label if you were an artist. Boy, everything has really changed since then. Now virtually anyone can have a pretty good sounding personal studio without spending a lot of money, record labels have lost much of their power, and since there aren't as many commercial studios around any more, there are fewer experienced engineers available to pass on the tricks of the trade. That's what makes this book all the more useful.

The idea behind my books is to preserve the techniques of the recording masters for history, and pass those techniques on to you. That might not be as hands-on or efficient as the engineer/assistant system used in large studio facilities for more than fifty years, but at least there's somewhere to refer to if you don't know how to record an instrument and there's no one around to ask. Yes, you can always turn to Google, but you're just as likely to get some questionable information as you are to get pearls of wisdom.

In this new era of samples, loops and modeling, a whole generation of engineers have grown up with minimal working knowledge of microphone technique, and that's understandable when you can make great recordings without ever having to do much tracking in the first place. The problem is that sooner or later there'll come a time when a question like "What's the best way to mike the snare to really make it punchy?", "How do I get a big guitar sound like (name your favorite artist) gets?", or "How do you mike a piccolo?" can cause a mild panic. That's where this book comes in.

While there are many books available that touch upon the basics of recording (especially stereo orchestral material), there aren't many books that feature multiple techniques in miking a wide variety of instruments in the detail needed to achieve a reasonable and consistent result. And there is no book that concentrates upon this basic, yet all-important facet of recording in quite the same way as it's presented here.

As you'll see, there are many ways to get the same basic result. There's no right way to mic an instrument, but some ways are more accepted than others, and therefore they become a "standard." Whenever possible, I've tried to provide a high resolution photo or diagram of a described miking technique, as well as a written description of the theory behind it, as well as the possible variables.

For those of you who have read my previous books, you'll notice that the format for this book is identical to those. It's divided into two sections;

Part 1 takes a look at the microphone basics, the classic models frequently used, and the techniques used by the best tracking engineers in the business.

And Part 2 is comprised of interviews with some of the finest (and in some cases legendary) tracking engineers in the world.

Keep in mind that whenever possible, I tried not to get too specific on the make and model of microphone to use. That's because you probably don't have access to some of the expensive vintage mics that are frequently referred to in the various setups, but the fact of the matter is that it's the placement that counts more than the mic, so feel free to use whatever mics you have.

Meet The Engineers

Here's a list of the engineers who contributed to this book along with some of their credits. You'll find that there are some industry legends here as well as others that specialize in all different genres of music. One big change in this edition. I've moved many of the interviews that appeared in the previous editions over to my website. You can access the full interviews (not just the edited ones that made the book) at: BobbyOwsinski.com/interviews.

As you read each engineer's accomplishments and you see that are distinctly "old school," don't be fooled into thinking that the information they're imparting is not applicable today. Their techniques are evergreen and work just as well now as it did during their respective heydays.

Chuck Ainlay is one of the new breed of Nashville engineers that brings a rock and roll approach to country music sensibility. With credits like George Strait, Dixie Chicks, Vince Gill, Patty Loveless, Wynonna and even such rock icons like Dire Straits and Mark Knopfler, Chuck's work is heard world-wide.

Grammy-winner **J.J. Blair** has worked with a variety of artists that include Rod Stewart, Johnny Cash, June Carter Cash, Jeff Beck, George Benson, P. Diddy, Smokey Robinson, and many more. I've worked with a lot of great engineers, but only J.J. has that combination of musicianship (he's an excellent guitarist), engineering skill, audio electronics knowledge, and attention to detail that makes him totally unique.

The late **Michael Bishop** was formerly the chief engineer for the audiophile label Telarc, who mostly utilized the "old school" method of mixing live on the fly, and always with spectacular results.

Bruce Botnick has a perspective on recording that few engineers have. After starting his career in the thick of the Los Angeles rock scene, Bruce became one of the most in-demand movie soundtrack recordists and mixers with a long list of blockbuster credits.

Wyn Davis is best known for his work with hard rock bands like Dio, Dokken and Great White, as well as a host of live TV concerts. From his Total Access studios in Redondo Beach, California, Wyn's work typifies old-school engineering coupled with the best of modern techniques.

Eddie Kramer is unquestionably one of the most renowned and well-respected producer/engineers in all of rock history. His credits include rock icons such as Jimi Hendrix, The Beatles, The Rolling Stones, Led Zeppelin, Kiss, Traffic and The Kinks, to Pop stars Sammy Davis Jr. and Petula Clark, as well as recording the seminal rock movie "Woodstock."

Mack has a Who's Who list of credits such as Queen, Led Zeppelin, Deep Purple, The Rolling Stones, Black Sabbath, Electric Light Orchestra, and many more. Having recorded so many big hits that have become the fabric of our listening history, Mack's engineering approach is steeped in European classical technique coupled with just the right amount of rock & roll attitude.

Grammy-winner **Sylvia Massy** has worked with a wide variety of artists across many different musical genres. These include Tool, Johnny Cash, Red Hot Chili Peppers, System of a Down, Prince, Sheryl Crow, Tom Petty, Smashing Pumpkins, and Lenny Kravitz, to name just a few. She is also a co-owner of the largest microphone museum in the world, with over 2,200 vintage pieces, including many rare prototypes.

Known as "the drummer's engineer" for his smooth, natural drum sound, **Dennis Moody** has been the choice of top-shelf session drummers like Dave Weckl, Steve Gadd, Michael White, and the late great Ricky Lawson when it comes to their own personal recording projects. Dennis is also one of the few studio engineers who regularly crosses over to live concert engineering, having mixed Front Of House in venues like Carnegie Hall, Madison Square Garden, Wembley Arena, and Royal Albert Hall.

Barry Rudolph has worked with many music legends and is credited with 30 gold and platinum albums. What makes him unique is that he's written over 6,000 audio gear reviews for Mix Magazine, Music Connection and Resolution and others over the years, so he's played with more gear than most people ever will in several lifetimes.

Al Schmitt has won more engineering Grammys than anyone in history, with 20 on his mantle. Couple that with his work on over 150 gold and platinum records, Al's credit list is way too long to print here (Henry Mancini, Steely Dan, Neil Young, Bob Dylan, Paul McCartney, George Benson, Toto, Natalie Cole, Quincy Jones, and Diana Krall are some of them), but suffice it to say that his name is synonymous with the highest art that recording has to offer. Al's ears remained golden as he continued to packed recording schedule right up until his passing in 2021.

Also included are interviews with these special non-engineer guests:

David Bock knows as much about microphones anyone on the planet. From repairing vintage mikes of all kinds to building newer versions of the classics, David knows why and how they work, and why they're made the way they are.

Ross Garfield is known as "The Drum Doctor," and anyone recording in Los Angeles certainly knows that he's the guy to either rent a great sounding kit from, or have him fine-tune your kit. Having made the drums sound great on platinum selling recordings for the likes of Bruce Springsteen, Rod Stewart,

Metallica, Marilyn Manson, Dwight Yokum, Jane's Addiction, Red Hot Chili Peppers, Foo Fighters, Lenny Kravitz, Michael Jackson, Sheryl Crow, and many more than what can comfortably fit on this page, Ross agreed to share his insights on what it takes to make drums sound special.

I'm sure you'll find these interviews as much fun to read as they were for me to conduct. I'm also sure that even if you're pretty good at recording already, you'll find some interesting techniques in the book that you never thought of and might find useful along the way. I know I did.

By the way, you'll also find additional interviews from previous versions of this book at bobbyowsinski.com/interviews.

PART I

RECORDING

HOW MICROPHONES WORK

Microphones appear in an almost endless variety of shapes, sizes, and design types, but no matter what their physical attributes, their purpose is same; to convert acoustic vibrations in the form of air pressure into electrical energy so they can be amplified or recorded. Most achieve this by the action of the air vibrating a diaphragm connected to another component that either creates or allows a small electron flow.

There are three basic mechanical techniques that are used in building microphones for professional audio purposes, but all three types have the same three major parts:

A Diaphragm: Sound waves strike the diaphragm, causing it to vibrate. To accurately reproduce high frequency sounds, it must be as light as possible.

A Transducer: The mechanical vibrations of the diaphragm are converted into an electronic signal by the transducer.

A Casing: As well as providing mechanical support and protection for the diaphragm and transducer, the casing can also be made to help control the directional response of the microphone.

Let's take a closer look at the three types of microphones.

THE DYNAMIC MICROPHONE

The dynamic microphone is the workhorse of the microphone family. Ranging from fairly inexpensive to moderately expensive, there's a model of dynamic mic available to fit just about any application.

How It Works

In a moving-coil (or more commonly called *dynamic*) microphone, sound waves cause movement of a thin metallic diaphragm and an attached coil of wire that is located inside a permanent magnet. When sound waves make the diaphragm vibrate, the connected coils also vibrate within the magnetic field, causing current to flow because of what's known as *electromagnetic induction*. Because the current is produced by the motion of the diaphragm and the amount of current is determined by the speed of that motion, this kind of microphone is also known as *velocity sensitive* (see Figure 1.1).

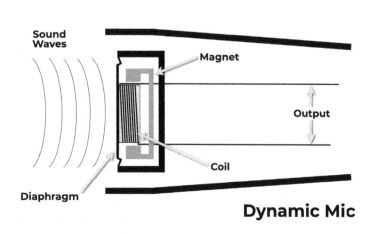

Figure 1.1: Dynamic mic block diagram
© 2023 Bobby Owsinski

The ability of the microphone to respond to transients and higher frequency signals depends upon how heavy the moving parts are. In this type of microphone, both the diaphragm and the coil move, so that means its moving parts are relatively heavy, more from the coil than anything else. As a result, the frequency response falls off above about 10kHz because it just can't respond quickly enough to reproduce the higher frequencies due to the weight of the coil and diaphragm.

The microphone also has a resonant frequency (a frequency or group of frequencies that is emphasized) that is typically somewhere from about 1kHz to 4kHz. This resonant response is sometimes called the *presence peak*, because it occurs in the frequency region that directly affects voice intelligibility. Because of this naturally occurring effect, dynamic microphones are often preferred by vocalists, especially in live sound applications.

Dynamic microphones with extended frequency range and a flat response tend to be moderately expensive because they're somewhat complex to manufacture with the required precision, but they're generally very robust (you can actually hammer nails with some of them and they'll still work!) and insensitive to changes in humidity.

Table 1.1 Dynamic Microphone Advantages And Disadvantages

DYNAMIC MIC ADVANTAGES	DYNAMIC MIC DISADVANTAGES
Robust and durable	Resonant peak in the frequency response
Can be relatively inexpensive	Typically weak high-frequency response beyond 10kHz or so
Insensitive to changes in humidity	
Needs no external or internal power to operate	
Can be made fairly small	
Handles high sound pressure levels well	

THE RIBBON MICROPHONE

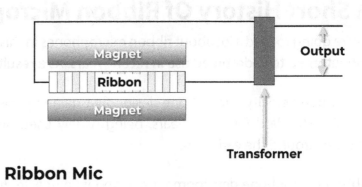

Ribbon Mic

Figure 1.2: Ribbon mic block diagram
© 2023 Bobby Owsinski

The ribbon microphone operates using almost the same principle as the moving-coil microphone. The major difference is that the transducer is a strip of extremely thin aluminum foil wide enough and light enough to be vibrated directly by the moving molecules of air of the sound wave, so no separate diaphragm is necessary. However, the electrical signal generated is very small compared to that of a moving-coil microphone, so an output transformer is needed to boost the signal to a usable level (see Figures 1.2 and 1.3).

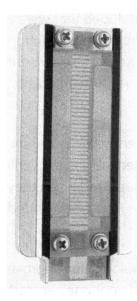

Figure 1.3: Ribbon mic transducer
Courtesy of Royer Labs

As with the dynamic microphone, the high-frequency response is governed by the mass of the moving parts, but because the diaphragm is also the transducer, the mass is usually a lot less than a dynamic type. As a result, the upper frequency response tends to reach higher than that of a dynamic mic, typically up to around 14kHz. The frequency response is also generally flatter than for a moving-coil microphone.

All good studio ribbon mics provide more opportunity to EQ to taste because they "take" EQ well. Ribbon mics have their resonance peak more toward the bottom of their frequency range, which means that a ribbon doesn't inherently add any extra high-frequency hype like condenser mics do.

Table 1.2: Ribbon Microphone Advantages And Disadvantages

RIBBON MIC ADVANTAGES	RIBBON MIC DISADVANTAGES
Relatively flat frequency response	Fragile; requires care during operation and handling
Extended high-frequency response as compared to dynamics	Moderately expensive
Needs no external or internal power to operate	The ribbon element may be stressed if 48-volt phantom power is continuously applied

A Short History Of Ribbon Microphones

You're going to read a lot about ribbon microphones in this book because they have been rediscovered and returned to widespread use in recent years. As a result, a bit of history seems in order.

The ribbon-velocity microphone design first gained popularity in the early 1930s and remained the industry standard for many years, being widely used on recordings and broadcasts from the '30s through around the early '60s.

Ribbon microphone development reached its pinnacle during this period. Though they were always popular with announcers and considered state-of-the-art at the time, one of the major disadvantages of early ribbon mics was their large size, since the magnet structures and transformers of the time were bulky and inefficient. When television gained popularity in the late 1940s, their size made them intrusive on camera and difficult to maneuver, so broadcasters soon looked for a replacement more suitable to the television on-camera environment.

About that time, a newer breed of condenser and dynamic microphones were developed that were a lot more compact and far more rugged. As a result, television and radio began to replace their ribbon mics with these new designs. Since ribbon mics were being used less and less, further development was considered unnecessary, and the mic soon suffered a fate similar to that of the vacuum tube when transistors hit the market.

Although ribbon mics might have been out of favor in broadcast, recording engineers never quite gave up on the technology. While always fragile because of the thin ribbon material used, ribbon mics still provided some of the sweetest sounds in recording, as most old-school engineers realized when they A/B'd them against the latest microphone technology. As a result, vintage ribbon mics commanded extremely high prices in the used marketplace.

Recently, a few modern manufacturers have begun to not only revive the technology, but improve it as well. Companies such as Coles, Beyer, Royer, and AEA now make ribbon microphones at least as good as, if not better than, the originals, and they are a lot more robust as well. Thanks to recent developments in magnetics, electronics, and mechanical construction, modern ribbon microphones can be produced smaller and lighter yet still maintain the sound of their vintage forbearers while achieving sensitivity levels matching those of other types of modern microphones. Their smooth frequency response, ability to handle higher sound pressure levels, and phase linearity make them ideally suited for the digital formats that dominate the industry today.

THE CONDENSER MICROPHONE

The condenser microphone has two electrically charged plates; one that can move, which acts as a diaphragm, and one that's fixed, called a *backplate*. This is, in effect, a capacitor (also known as a *condenser*) with both positively and negatively charged electrodes and an air space in between. Sound waves depress the diaphragm, causing a change in the spacing between it and the backplate, which in turn changes the capacitance. This change in capacitance and distance between it and the backplate causes a change in voltage potential that can be amplified to a usable level.

To boost this extremely small voltage, a vacuum tube or transistor amplifier is incorporated into the mic itself. This is why a battery or external power (called phantom power) is required, because power is needed to charge the plates and also to run the preamp. Because the voltage requirements to power a vacuum tube are high (usually between 100 and 200 volts) and therefore require some large and heavy components, some condenser microphones have their power supply in a separate outboard box (see Figure 1.4).

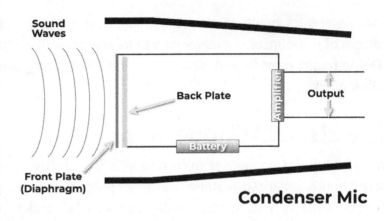

Figure 1.4: Condenser mic block diagram
© 2023 Bobby Owsinski

Condenser microphones come in two types: large diaphragm and small diaphragm. According to premier condenser microphone manufacturer Neumann, "Historically, large diaphragm condensers came first. Early condenser microphones of the 1930s and 40s had to use large diaphragm capsules to overcome the noise of the tube electronics. A large membrane captures more acoustic energy, thus generating a higher signal voltage. Small diaphragm condenser microphones with a decent signal-to-noise ratio only became feasible when dedicated microphone tubes and low noise transistors became available in the 1950s and 60s."

The biggest advantage of a large diaphragm condenser microphone is still the noise performance, which is typically around 6dB lower than even a state-of-the-art small diaphragm condenser. That's where the advantages stop, however. A small diaphragm condenser has better transient response (it reacts faster to the sound waves), has a higher frequency response, and has a more consistent pickup pattern.

Speaking of pickup pattern, a condenser has an omnidirectional pickup pattern in its native state. To make it directional, little holes are punched in the backplate. The object of the holes is to delay the arrival of sound at the rear of the diaphragm to coincide with the same sound at the front, which then cancels out the sound. The size and position of the holes determines the frequencies that will be cancelled.

Most large-diaphragm condensers are multi-pattern microphones. This design is composed of a single backplate placed between two diaphragms. By varying how much signal from each diaphragm is fed to the preamp, the microphone can have selectable patterns ranging from a tight cardioid to a figure-eight to full omnidirectional, which is why the pattern control on older tube mics is continuously variable.

Condenser mics, however, always resonate a bit, typically in the 8k to 12kHz range. A condenser mic's pattern of resonances is a major part of its character. Their built-in top-end response bump limits the EQ you might want to add, since a little bit of high-frequency boost can start to sound a bit "edgy" rather quickly.

The Electret Condenser

Another type of condenser microphone is the electret condenser. An electret microphone uses a permanently polarized material (called *electret*) as a diaphragm, thus avoiding the necessity for the external voltage that's required in a conventional condenser. Electrets can be made very small and inexpensively and are the typical microphones found on items such as cell phones, laptops, video recorders and portable audio recorders.

Better-quality electret condensers incorporate a preamplifier to match their extremely high impedance and boost the signal. One of the problems with many of the early electret condenser microphones is that the electret material loses its charge over time, causing the output and frequency response to diminish, but this is less of a concern with modern versions.

Table 1.3: Condenser Microphone Advantages And Disadvantages

CONDENSER MIC ADVANTAGES	CONDENSER MIC DISADVANTAGES
Excellent high frequency and upper harmonic response	Moderate to very expensive
Can have excellent low-frequency response	Requires power (either internal or external)
Excellent transient response	Can be relatively bulky
Can have changeable polar patterns	Low-cost models can suffer from poor or inconsistent frequency response
	Two mics of the same model may sound somewhat different
	Humidity and temperature affect performance

Condenser Mic Fallacies

Here are a number of popular misconceptions about condenser microphones, along with the explanations as to why they're not true.

- ***A large-diaphragm condenser has better low-frequency response than a small-diaphragm condenser.*** This is not necessarily true. In many cases, small-diaphragm condensers reproduce the low end just as well as their larger kin.

- ***A cardioid condenser has a better low-frequency response than an omni.*** Not true. In condenser mics with an omnidirectional polar response, the bass response is only limited by the electronics. Even a very small-diaphragm mic can have a flat response down to below 20Hz.

- ***A large-diaphragm condenser has a flatter response than a small-diaphragm condenser.*** Not true. Large-format capsules are prone to low-frequency resonance, which means that they can have trouble reproducing low frequencies at a high level. They can also "bottom out" as a result of the diaphragm hitting the backplate, which is the popping that can occur when a singer is too close to a microphone without a pop filter. To minimize this, some microphones over-damp the capsule, making the mic sound either thin or alternatively lumpy in response, while some address this by adding a low-frequency roll-off or EQ circuitry to try to put back frequencies suppressed in the capsule.

- ***A small-diaphragm condenser is quieter than a large-diaphragm.*** Not true. The difference in the size of the diaphragm translates into a difference in signal-to-noise ratio. The bigger diaphragm provides more signal for a certain electrical noise level and therefore can be quieter than the small diaphragm.

- ***Condenser mics have consistent response from mic to mic.*** They're not as close as you might think. Despite what the specs might say, there can be vast differences in the sound between two mics of the same model, especially in the less expensive models. This particularly applies to tube-type mics, where there are not only differences between the capsules, but also matching of the tubes.

Unless two mics are specifically matched in their frequency response, differences between them are inevitable. That said, the value of precise matching of microphones is open to much debate. One school of thought says that you need closely matched response for a more precise stereo soundfield, while another school thinks that the difference can actually enhance the soundfield.

Condenser Microphone Operational Hints

Condenser mics can sometimes require some extra attention. Here are a number of tips that can not only prolong the life of your mic, but also keep its performance as high as the day it left the factory.

- **The most commonly seen problem with condenser microphones is dirt on the capsule, which causes the high-end response to fall off.** Since a condenser is always carrying a static charge when operating, it will automatically attract small airborne particles. Add to this people singing and breathing into it, and you have a frequency response that's slowly deteriorating. Because the metal film of the capsule is very thin, the layer of dirt can actually be much thicker than the original metal film and polymer support. *Despite what is commonly believed, the mesh grill of the mic will not do much more than stop people or objects from touching the capsule, and the acoustic foam inside the grill has limited effect.*

- **Cleaning a capsule is a very delicate and potentially damaging operation that is best left to a professional,** so the next best thing is preventive maintenance.

1. Always use an external pop filter.
2. Keep your condenser microphones cased when not in use.
3. Cover the mic if it will be left set up on a stand overnight.

- **Humidity and temperature extremes can have undesirable effects on performance.** When exposed to a warm or humid room after a period of very low temperature, condensation in the casing can cause unwanted noises or even no signal output until the unit has dried out.

- **Don't blow into the microphone.** Some diaphragms may bottom out and hit the backplate and then stick there. Switching off the microphone and disconnecting the power supply may unstick it, though.

- **A condenser microphone can be overloaded, which can cause either distortion or harshness of tone.** Usually this is not from the diaphragm overloading but due to the high output from the capsule overloading the built-in preamplifier. This is less likely to happen in the case of a vacuum-tube model since tubes naturally overload in a sonically unobtrusive manner (sometimes called *soft clip*). Most internal mic preamps have a -10dB pad switch to lower the output from the capsule. In the event that this amount is still insufficient, the bottom-end roll-off filter will also reduce power from the capsule.

Phantom Power

Unlike dynamic and ribbon microphones, all condenser microphones require power of some type. Older tube condensers require an outboard power supply, while electret condensers are sometimes powered with a battery. All other condenser microphones require power from an outside source called *phantom power*. This is a 48-volt DC power source fed by a recording console, microphone preamp, or DAW interface over the same cable that carries the audio. On most recording consoles phantom power is switchable, since it may destroy the internal ribbon on many older ribbon mics. It may cause a loud pop when disconnecting a cable connected to a dynamic mic as well.

THE MEMS MICROPHONE

The latest advancement in microphones is the solid-state MEMS, which stands for Micro-Electro-Mechanical System. A MEMS microphone is an electro-acoustic transducer that combines a sensor (MEMS) and an application-specific integrated circuit (ASIC) into a single package. The sensor converts the sound pressure to a capacitance change like a condenser microphone, and the ASIC manages the MEMS polarization and the analog and digital outputs.

One of the biggest advantages of a MEMS is that it can contain an analog to digital convertor right on the same chip so the output can be fed directly into your computer without the need of a computer audio interface.

MEMS mics are already being used in many devices that you probably own, from smartwatches and smart speakers, to phones, noise canceling headphones and laptops. The Zylia ZM-1 (see Chapter

12) and the Fender Stratacoustic are examples where the technology is used in a professional audio environment, although more examples will soon be available from other manufacturers.

Table 1.4: MEMS Microphone Advantages And Disadvantages

MEMS MIC ADVANTAGES	MEMS MIC DISADVANTAGES
Much less expensive than other types of microphones	Needs power to operate, although the requirements are very low
Virtually identical performance from mic to mic	
Lower noise than condenser microphones	
Can work in high SPL environments	
Extremely small and lightweight	

MICROPHONE SPECIFICATIONS

While hardly anyone selects a microphone solely on specifications, it's good to be clear on the various parameters involved. The following won't delve too much into the actual electronic specs as much as how those specs apply for your application.

Sensitivity

This is a measure of how much output signal is produced by a given sound pressure. In other words, this tells you how much output you'll get from a microphone. Generally speaking, for the same sound pressure, ribbon microphones are the quietest while condensers, thanks to their built-in preamplifiers, provide the most output level.

Where this might be a concern is in how your signal chain responds when recording loud signals. For instance, a condenser mic placed on a loud source like a snare drum might easily overload the console preamp, outboard microphone preamp, or DAW interface because of the mic's inherent high output.

On the other hand, the low output of a ribbon mic placed on a quiet source might cause you to turn up that same mic preamp to such a point that electronic noise becomes an factor.

Sensitivity ratings for microphones may not be exactly comparable, since different manufacturers use different rating systems. Typically, the microphone output (in a soundfield of specified intensity) is stated in dB (decibels) compared to a reference level. Most reference levels are well above the output level of the microphone, so the resulting number in dB will be negative. Thus, as in Table 1.5, a condenser microphone with a sensitivity rating of -38 will provide a 16dB hotter signal than a microphone with a sensitivity of -54dB, which will in turn provide a 6dB hotter signal than one rated at -60dB. In other words, the closer to 0dB the sensitivity is, the higher the mic's output will be. *Note that good sensitivity does not necessarily make a microphone "better" for an application.*

Table 1.5: Microphone Sensitivity chart

RIBBON SENSITIVITY	DYNAMIC SENSITIVITY	CONDENSER SENSITIVITY
- 60 (Beyer M 160)	- 54 (Shure SM57)	- 38 (Neumann U 87 in omni)

© 2023 Bobby Owsinski

Overload Characteristics

Any microphone will produce distortion when it's overdriven by a loud sound level. This is caused by various factors. With a dynamic microphone, the coil may be pulled out of the magnetic field. In a condenser, the internal amplifier might distort. Sustained overdriving or extremely loud sounds can permanently distort the diaphragm, degrading performance at ordinary sound levels. In the case of a ribbon mic, the ribbon could be stretched out of shape, again causing the performance to seriously degrade.

Loud sound levels are encountered more often than you might think, especially if you place the mic very close to loud instruments like a snare drum or the bell of a trumpet. In fact, in many large studio facilities, a microphone that has been used on a kick drum or snare, for instance, is labeled as such and is not used on any other instrument afterward.

Frequency Response

Although a flat frequency response has been the main goal of microphone companies for the last three or four decades, that doesn't necessarily mean that a mic with a flat response is the right one for the job. In fact, a "colored" microphone can be more desirable in some applications where the source has either too much emphasis in a frequency range or not enough. Many mics have a deliberate emphasis at certain frequencies, which makes them useful for some applications (vocals in a live on-stage situation, for example). In general, though, problems in frequency response are mostly encountered with sounds originating off-axis from the mic's principal directional pattern (see Directional Response below).

Free-Field or Diffuse-Field

Free-field means that most of the sound that the mic hears comes from the source. Diffuse-field means that the room reflections play a large role in what the mic hears.

Mics designed for free-field use usually have a somewhat flat frequency response in the high frequencies, and as a result can sound dull when placed farther away from the source in the room. Mics designed for the diffuse-field have a boost in the upper frequencies that make them sound flat when placed farther away, but it can make them sound too bright if used for close-miking a source. The Neumann M50 is an example of a diffuse-field microphone (see Chapter 6 under Decca Tree).

Noise

Noise in a microphone comes in two varieties: self-noise generated by the mic itself (like in the case of condenser microphones) and handling noise.

Condenser microphones are most prone to self-noise because a preamplifier must be used to amplify the minuscule signal that's produced by the capsule. Indeed, the audio signal level must be amplified by a factor of over a thousand, so any electrical noise produced by the microphone will also be amplified by that amount, making even slight amounts of noise intolerable. Dynamic and ribbon microphones are essentially noise free, but they are subject to handling noise.

Handling noise is the unwanted pickup of mechanical vibration through the body of the microphone. Many microphones intended for handheld use require very sophisticated shock mountings built inside the shell of the mic to reduce the handling noise.

Directional Response

The directional response of a microphone is the way in which the microphone responds to sounds coming from different directions around the microphone. The directional response is determined more by the casing surrounding the microphone than by the type of transducer it uses.

The directional response of a microphone is recorded on a polar diagram. This polar diagram shows the level of signal pickup (sometimes shown in decibels) from all angles and in different frequency ranges. *It should be noted that all mics respond differently at different frequencies.* For example, a mic can be very directional at one frequency (usually higher frequencies) but virtually omnidirectional at another.

A microphone's polar response pattern can determine its usefulness in different applications, particularly multi-microphone settings where proximity of sound sources makes microphone leakage a problem.

There are four typical patterns commonly found in microphone design.

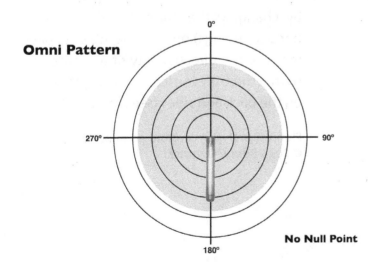

Omni-Directional

An omni-directional microphone picks up sound almost equally from all directions. The ideal omni-directional response is where equal pickup occurs from all directions at all frequencies (see Figure 1.5).

Figure 1.5: Omni-directional polar pattern
© 2023 Bobby Owsinski

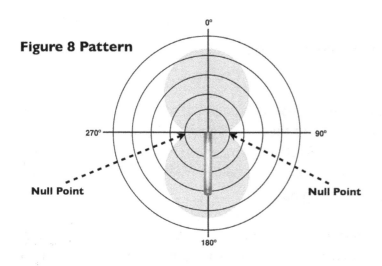

Figure 1.6: Figure 8 polar pattern.
© 2023 Bobby Owsinski

Figure of Eight

Figure of eight (sometimes called simply *figure 8* or *bi-directional*) microphones pick up almost equally in the front and back but nearly nothing on each side. It should be noted that the frequency response is usually a little better (as in brighter) on the front side of the microphone, although the level between front and rear can seem about the same.

Because the sensitivity on the sides is so low, figure 8s are often used when a high degree of rejection is required (see Figure 1.6).

Cardioid

The cardioid microphone has strong pickup on the axis (in the front) of the microphone but reduced pickup off-axis (to the side and to the back). This provides a more or less heart-shaped pattern, hence the name *cardioid* (see Figure 1.7).

Figure 1.7: Cardioid polar pattern.
© 2023 Bobby Owsinski

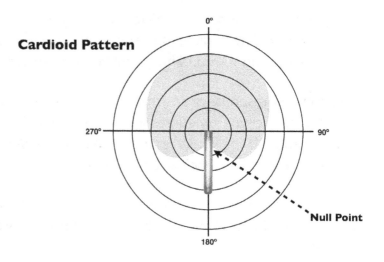

Hyper-Cardioid Microphones

By changing the number and size of the ports (openings) on the case, it's possible to increase the directionality of a microphone so that there is even less sensitivity to sounds on the back and sides (see Figure 1.8).

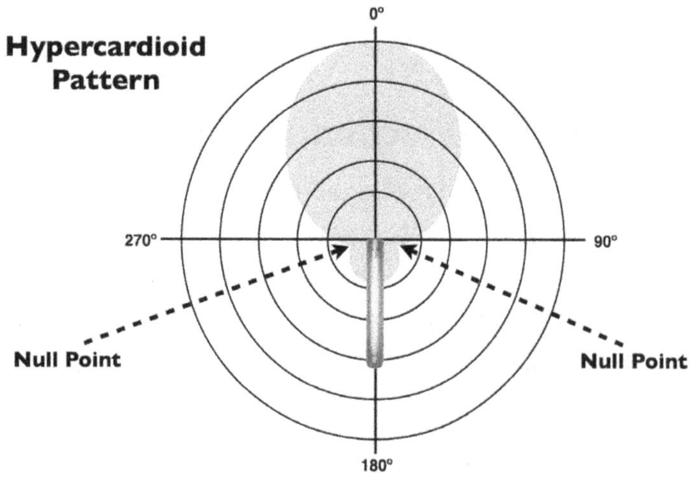

Figure 1.8: Hyper-cardioid polar pattern.
© 2023 Bobby Owsinski

Proximity Effect

A peculiarity of the microphones with a directional pattern is that they have a different frequency response when placed close to or far away from the sound source. Cardioid and hyper-cardioid microphones experience low-frequency build-up the closer you place to the mic, which is known as proximity effect. In many cases this can be used to good effect, adding warmth and fullness to the source signal, but it can also make the frequency response seem out of balance if it's not taken into account.

Since proximity effect occurs on frequencies below 200Hz, one place where you can hear this most noticeably is on the 80Hz low E of an acoustic guitar when the microphone is placed too close, causing it to sound boomy.

SPECIALTY MICROPHONES

While the vast majority of microphones manufactured are general purpose and have a wide variety of uses, there are some mics that serve a single particular function. Let's look at them.

Shotgun Microphones

Figure 1.9: Sennheiser MKH416 shotgun mic
Courtesy of Sennheiser USA.

There are a number of applications that require an extremely directional microphone, such as in news gathering, wildlife recording, or recording dialogue on movie and television sets. One such microphone is the shotgun (occasionally called the "rifle", or the more technical "interference tube") microphone. This mic consists of long tube with slots cut in it, connected to a cardioid microphone (see Figure 1.9).

Sound arriving from the sides enters through the slots in the tube, which causes some of the sound to cancel at the diaphragm or capsule. Sound entering at the end of the tube goes directly to the microphone capsule, providing a large difference between the source and other background noise. The tube is normally covered with a furry windshield for outdoor use.

Lavaliere Microphones

Extremely small "tie clip" microphones are known as lavaliere mics (sometimes just called lavs) and are made for situations where a handheld or mounted microphone is not appropriate. Lavs are usually electret condenser and omnidirectional and are designed to blend in with an article of clothing. One of the major problems with lavalieres is handling noise, which can be quite severe if an article of clothing (such as a jacket) is rubbing against it. Therefore, placement on the body becomes crucial (see Figure 1.10).

Figure 1.10: Rode SmartLav lavaliere mic
Courtesy of Sennheiser USA

TIP: There are many ways to place a lav, especially during a theatrical or movie production, but the standard position is about the height of a normal chest pocket on a suit jacket, or right above the heart where the chest caves in a little bit. If you place it higher you get more clothes rustle, and if you place it close to the neck you lose a lot of the high frequencies, as the chin shades the direct sound.

PZM Microphones

The pressure-zone microphone, or PZM, is designed to limit the amount of phase coloration from the early reflections from a sound source in a room. It accomplishes this by placing the microphone capsule very close to a flat surface. This flat surface is called the boundary and is why this type of microphone is also called a "boundary microphone."

Figure 1.11: Crown PZM-30D microphone
Courtesy of Crown Audio, Inc.

By getting the microphone capsule close to the boundary, it cuts down on the large number of reflected sound waves hitting it from all angles. The waves that are reflected off the closely positioned boundary are much stronger than waves that have bounced all around the room. This helps the microphone to become more sensitive, and as a result keeps the audio from sounding too reverberant.

PZM microphones, which are omnidirectional, are flat and designed to be mounted to a wall or placed on the floor or a tabletop. The bigger the boundary underneath the microphone, the better it will perform (see Figure 1.11).

Wireless Microphones

It's long been the dream of many performers to increase their freedom by removing the connecting cable from the microphone, and guitarists in the studio have wanted to play in the control room ever since overdubs became possible. Until recently, wireless systems weren't of sufficient quality to use in the studio, but the latest generation now rivals the wired versions.

Wireless Microphone Components

A wireless system consists of three main components: an input device, a transmitter, and a receiver. The input device provides the audio signal that will be sent out by the transmitter. It may be a microphone, such as a handheld vocalist's model or a lavaliere "tie-clip" type. With wireless systems designed for use with electric guitars, the guitar itself is the input device.

The transmitter handles the conversion of the audio signal into a radio signal and broadcasts it through an antenna. The antenna may stick out from the bottom of the transmitter or it may be concealed inside. The strength of the radio signal is limited by government regulations. The distance that the signal can effectively travel ranges from 100 feet to more than 1,000 feet, depending on conditions.

Figure 1.12: Shure ULXD4 wireless transmitter and receiver
Courtesy of Shure USA

Transmitters are available in two basic types. One type, called a body-pack or belt-pack transmitter, is a small box about the same size as cell phone. The transmitter clips to the user's belt or may be worn on the body. For instrument applications, a body-pack transmitter is often clipped to a guitar strap or attached directly to an instrument such as a trumpet or saxophone. In the case of a handheld wireless microphone, the transmitter is built into the handle of the microphone, resulting in a wireless mic that is only slightly larger than a standard wired microphone. Usually, a variety of microphone elements or "heads" are available for handheld wireless microphones. All wireless transmitters require a battery (usually a 9-volt alkaline type or internal rechargeable) to operate (see Figure 1.12).

The job of the receiver is to pick up the radio signal broadcast by the transmitter and change it back into an audio signal. The output of the receiver is electrically identical to a standard microphone signal and can be connected to a typical microphone input on a console, mic preamp, or DAW interface.

Wireless receivers are available in two different configurations. Single-antenna receivers utilize one receiving antenna and one tuner, similar to an FM radio. Single-antenna receivers work well in many applications but are sometimes subject to momentary interruptions or "dropouts" in the signal as the person holding or wearing the transmitter moves around the room.

A diversity-type receiver uses two separate antennas spaced a short distance apart, utilizing (usually) two separate tuners to provide better wireless microphone performance. A circuit in the receiver automatically selects the stronger of the two signals, or in some cases a blend of both. Since one of the antennas will almost certainly be receiving a clean signal at any given moment, the chances of a dropout occurring are reduced.

Analog wireless systems operate in two different frequency spectrums: VHF and UHF. Audio performance for VHF and UHF is nearly identical, but some of the high-end (and much more expensive) UHF systems offer real improvements in audio bandwidth, transient response, and system noise floor. In terms of operational range or distance, UHF offers some advantage especially in inhospitable RF environments. Another advantage is that broadband RF interference (compressors, elevator motors, computers, and so on) are often below UHF frequencies.

The preferred option for a wireless mic package these days is the digital type, which transmits a digital audio signal on a UHF bandwidth instead of an analog signal. The advantage is that the signal can have a flatter frequency response with greater dynamic range and less noise and interference than a comparable analog UHF system, plus have about 40% more battery life as well. An added advantage is that the signal can be digitally encrypted, making it can be virtually impossible for someone outside the venue with a scanning receiver to eavesdrop on the transmission.

Stereo Microphones

Stereo microphones are essentially two microphone capsules in a single casing or body. These are designed primarily for ease of placement, since a single stereo mic body is considerably smaller than two separate microphones and their mic stands and cables. An added advantage is that the capsules are normally closely matched in frequency and sensitivity response. Many times the capsules also rotate in order to provide a narrow or widened soundfield. Examples of stereo mics are the Royer SF-12, Neumann SM 69, Shure VP88, and AKG C 24 (see Figure 1.13).

Figure 1.13: Royer SF-24 stereo mic
Courtesy of Royer Labs

Parabolic Microphones

If you watch football on television, you've probably seen a parabolic microphone on the sidelines. This is usually a handheld dish that an operator will point at the field in an attempt to pick up some of the sounds of the game.

Figure 1.14: Klover Products Mik 26 parabolic mic
Courtesy of Klover Products

Similar to a radio telescope, a parabolic microphone is essentially an omnidirectional microphone that's pointed toward the middle of a rounded (parabolic) dish. The dish provides acoustical amplification by focusing the reflected sound onto one place. If the dish itself acoustically amplifies a certain frequency range by 10dB, it means that there's 10dB less electronic amplification (and therefore 10dB less noise) required within that range, so there will be less noise. The acoustical amplification increases with frequency, with the lowest frequency captured dependent upon the diameter of the dish.

The Recording Engineer's Handbook - 5th edition

The problem with parabolic mics is that they don't have a great low-frequency response, which would require a prohibitively large dish diameter. This tends to make them sound unnatural for many sounds unless the dish is really huge.

While widely used in sports broadcasting, it's not surprising that the parabolic microphone is one of the staples of the spying and espionage business as well. However, the most common use for parabolic mics in recording is to capture birdcalls, since most bird chirps are only composed of high frequencies (see Figure 1.14).

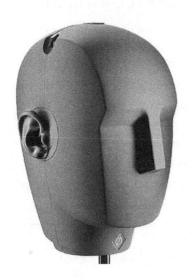

Binaural Microphones

Binaural recording is a method of capturing audio using two microphones, arranged about 7 inches apart and facing away from each other, to create a realistic 3D experience where the listener actually feel like he or she is in the room. This distance and placement roughly approximates the position of an average human's ear canals, but to truly maximize the experience, the mics are placed in a dummy head that more closely approximates how a human hears. The oldest examples of this type of microphone are the Neumann KU-80 and KU-100 (see Figure 1.15). See Chapter 12 for more on binaural and immersive recording.

Figure 1.15: Neumann KU-100 binaural microphone
Courtesy of Georg Neumann GmbH

MICROPHONE ACCESSORIES

A number of accessories are essential during normal microphone use in certain situations. Let's look at three of the most needed.

Pop Filters

Not to be confused with windscreens (see below), pop filters, either built into the mic (such as in a Shure SM58) or external, can either work great or be of little value. All microphones are subject to plosives or pops. However, many engineers are fooled into thinking that a foam windscreen is all that's needed to control them, when in fact positioning and microphone technique come more into play in the reduction of these "pops" (see Figure 1.16).

Figure 1.16: Samson SAPS 05 pop filter
Courtesy Samson Technologies

The problem with a pop screen built into a mic is that it's simply too close to the capsule to do much good in most cases. Wherever high-speed air meets an obstacle such as a pop screen, it will generate turbulence, which takes a few inches to dissipate. If the mic capsule is within that turbulence (as most are), it will still pop. Another problem with acoustic foam used within microphones is that it becomes brittle over time, and eventually little tiny bits of it break off and find their way inside the capsule (which is definitely not good for the sound quality).

Capturing spit from a vocalist is perhaps the best reason to use a pop screen. Condensation from breath can stop a vintage condenser microphone from working in a very short time.

External pop screens are designed to be as acoustically benign as possible, especially in the areas of transients and frequency response. That being said, they are not acoustically transparent, especially at very high frequencies. A U 87–style windscreen will knock the response down at 15kHz by 2 to 3dB, for instance.

Although there are many models of pop filters available commercially, it's fairly easy to build your own. Buy an embroidery hoop and some pantyhose, cut a leg of hose until you have roughly a square sheet, and clamp it in the embroidery hoop, then place it between the mic and singer.

TIP: Many people affix pop filters to a gooseneck device that attaches to the boom stand that holds the mic. It's usually easier to mount the pop filter on a second boom, as it makes positioning less frustrating and more exact.

Windscreens

Figure 1.17: A "blimp" windscreen (Source: iStock Photo)

Unlike breath pops, wind requires a completely different strategy. Wind isn't a nice, smooth flow of air, but rather turbulent and random. The noise that it causes is the change in air pressure physically moving the element or ribbon in the microphone. The vibration of wind (which is low-frequency in nature) against the element can be substantially stronger than the sound vibrations from the source. Also, the more turbulent the wind, the less you will be able to find the null in a directional microphone's response.

Although acoustic foam only may be sufficient for omni mics in gentle breezes, directional mics require more elaborate two-stage windscreens. For any amount of wind, a blimp, which is much more effective and will attenuate wind noise on the order of 20 to 30dB, is required. Companies like Lightwave and Rycote make a variety of blimps and windscreens that are frequently used for location recording (see Figure 1.17).

TIP: For windsceens in general, the larger it is, the more effective it will be. A spherical shape is best since it's the least affected from wind in all directions.

> **TIP:** *A trick that's often used on presidential outdoor speeches is to slip a condom (use the non-lubricated ones) over the microphone and then slip a foam windscreen over it so it's visually acceptable. Although the frequency response will suffer, the wind noise will be attenuated.*

Shock Mounts

Shock mounts are designed to shield the microphone from picking up transmission noises that occur through the mic stand. Shock mounting is largely dependent on the mass of the microphone. Large-diaphragm mics are larger in mass by nature and therefore present a greater sensitivity to mechanical noise. Small-diaphragm mics, on the other hand, have far less mass and therefore don't have the same sensitivity to handling noise of their larger cousins. As a result, the shock mount used on a larger mic has to be much looser, therefore causing the mount to be floppier and sometimes more difficult to position (see Figure 1.18).

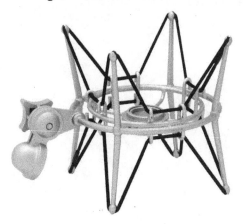

Figure 1.18: A shock mount.
© 2023 Bobby Owsinski

COMMON MICROPHONES

It used to be that every studio had virtually the same mics in their microphone lockers. Not so much because that's all that was available, but because that's what every engineer used. Today, there's a wide variety of mics that will do the job well, ranging from the all-time classics to new versions of the classics to some new mics that are fast becoming standards to even some inexpensive mics that do a great job despite their price. Let's take a look at each category.

THE CLASSIC MICROPHONES

A question I frequently used to get when I was teaching was what the microphones that I frequently talked about actually looked like. It's one thing to speak abstractly about placing a 47 FET on the kick or C12A's as overheads, but unless you know what they look like, you're totally in the dark. Likewise, this book discusses the use of various "classic" microphones, so it seemed appropriate to include a section with not only some pictures, but a bit of history as well.

Classic mics refer to the tried and true. Although their design may have come from the early days of recording or broadcast, over time they have proven that they can provide the sound that artists and engineers have found to be pleasing. While one of the goals of this book is to promote the theory that good technique and placement alone are sufficient in getting good sounds, having a few good microphones available is an important part of recording. Certainly, the following mics have proven to be successful tools over the long haul, and having one or more at your disposal will certainly help you in your quest for excellent-sounding recordings.

RCA 44 Ribbon Microphone

Developed in the late 1920s by the famous audio scientist Dr. Harry Olson, RCA's first permanent-magnet bidirectional ribbon microphone, the 44, entered the market in 1931. The 44 had a relatively low cost, which helped propel it to its legendary success and vast market penetration during that time.

The 44 series began with the 44-A, which was a relatively large microphone mostly because it used a large horseshoe magnet around the ribbon. The slightly larger 44-B was introduced in around 1938

with the 44-BX model soon after. All were bidirectional with a frequency response extending from 30Hz to 15kHz. In contrast to the 44-B, the 44-BX had the ribbon mounted more toward the rear of the case, which gave it a smaller figure-8 lobe on the back side. The 44-B was finished in a distinctive black with chrome ribbing on the lower portion, while the 44-BX was an umber gray and stainless steel. All versions of the mic featured two jumper positions within the case: V for voice, which substantially attenuated the low-frequency response, and M for music.

The 44-BX was manufactured up to around 1955. The 44-B/BX has become one of the classic influences in microphone technology, still in demand today, and has one of the most recognizable shapes in the world (see Figure 2.1).

Figure 2.1: RCA 44-BX

RCA 77 Unidirectional Ribbon Microphone

Realizing the need for a directional mic, Dr. Olson developed the unidirectional 77-A in the early 1930s. The 77-A, B, and C models utilized double ribbons to achieve the unidirectional pattern, and the C and D models were even capable of multiple patterns.

Improvements in magnet material allowed a significant reduction in size starting with the B model. The differences between the 77-D and 77-DX models are that the 77-DX had an improved magnet and transformer that produced a little more output. A screwdriver-operated switch was provided at the bottom of the lower shell with positions marked M for music and V1 and V2 for voice, where a high-pass filter was inserted into the circuit to attenuate the low frequencies. The 77 was discontinued around 1973, but its legacy continues as its shape remains the graphic icon for microphone that is recognized world-wide (see Figure 2.2).

Figure 2.2: RCA 77-DX
Source: iStock Photo

Neumann U 47

The original U 47, which was first marketed in 1948, was actually distributed by Telefunken. It was the first switchable-pattern condenser microphone, capable of both cardioid and omnidirectional patterns, and incorporated the highly successful 12-micron-thick M7 capsule and VF-14 tube amplifier (see Figure 2.3).

Figure 2.3: Neumann U 47
Courtesy of Neumann USA

The Recording Engineer's Handbook - 5th edition | 28

The U 47 was updated in 1956, when the capsule finish was changed from chrome to matte and the body length was reduced by about 3 inches. Also, the U 48, a cardioid/bidirectional version, was introduced that year. Two years later, Neumann's distribution deal with Telefunken ended, enabling Neumann to distribute their products under its own name.

Neumann U 47 FET

The U 47 FET started its life in 1969 as Neumann's answer to Sony and AKG's FET-based microphones (FET stands for *Field Effect Transistor*, which means the preamp was built around solid-state components instead of a tube). While originally designed to take the place of the tube U 47, the 47 FET never found acceptance in that role. Thanks to its fixed hyper-cardioid pattern and ability to take high SPL, the 47 FET eventually found a home in front of innumerable rock kick drums (see Figure 2.4).

Figure 2.4: Neumann U 47 FET
Courtesy of Neumann USA

Neumann U 67

With a streamlined, tapered body shape that has since become famous, Neumann introduced the U 67 in 1960. Thought of as an updated U 47, the U 67 featured a new Mylar film capsule, an internal 40Hz high-pass filter, and an amplifier pad switch to help overcome overload and proximity effect during close-up use. A three-way switch for selecting the directional pattern was added for extra versatility. The amplifier was based around the EF 86 tube (see Figure 2.5). Many pros consider this to be the most versatile microphone ever made, able to record just about anything well.

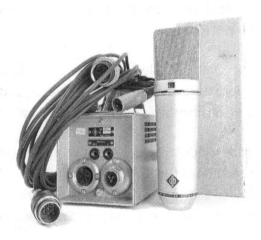

Figure 2.5: Neumann U 67
Courtesy of Neumann USA

Figure 2.6: Neumann M 49
Courtesy of Neumann USA

Neumann M 49/50

Designed in 1949, the M 49 was the first electronically remote-controlled variable-pattern condenser microphone (see Figure 2.6). The M 50, a lookalike twin of the M 49, shares the same design shape and the AC 701K tube, but it's strictly an omni pattern mic specifically designed for distant orchestral miking work. The mic features a high-frequency boost, and it becomes cardioid at high frequencies. The M 50 still reigns supreme as a Decca Tree microphone of choice for orchestral recording (see Chapter 6).

Neumann KM 84 Series

Figure 2.7: Neumann KM 84
Courtesy of Neumann USA

First introduced in 1966, the KM 84 was the first 48-volt phantom-powered microphone and one of the earliest FET mics. One of Neumann's all-time best-selling mics, it was made in the tens of thousands between 1966 and 1988. The KM 84 has a cardioid pickup pattern, while the KM 83 is omni and the KM 85 is cardioid with a fixed high-pass filter that rolls off at 4dB per octave beginning at 500Hz and reaches 12dB at 50Hz (see Figure 2.7).

In 1988, Neumann introduced the KM 100 series to replace the KM 80 series and incorporated several technical changes. In this series, the mics are modular with the FET amplifier in the capsule and not in the body of the mic itself. This enables the KM 100 series to have an extremely low profile (important for television work) since the mic body need not be directly attached to the capsule and can be located some distance away. The capsules are also interchangeable, with the AK 30 being omni and AK 40 cardioid. Thus, the KM 140 is the cardioid mic from the KM 100 series and is the direct descendant of the KM 84.

This AK 40 capsule was retuned just slightly from the original KM 64/84 in that a bump in the upper mids (approximately +4dB at 9kHz) was added. The self-noise, output level, and maximum SPL specifications were all improved over the older 84 as well.

Since modularity is expensive and engineers and musicians with project or home studios could not often afford the KM 140 as a result, the KM 184 was born. The same capsule was used from the KM 140, as well as the same FET, transformerless circuit, making the specs and performance the same. The KM 184 does not have a pad, and the capsules are not interchangeable.

Neumann KM 54/56

The KM 56 is a small-diaphragm tube condenser using an AC 701 tube and featuring a dual-diaphragm nickel capsule with three polar patterns (omni, figure-8, and cardioid) that are selectable on the body (see Figure 2.8).

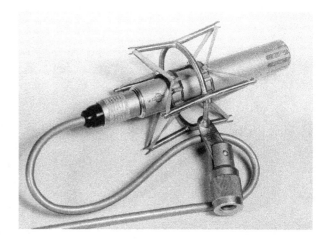

Figure 2.8: Neumann KM 54
Courtesy of Neumann USA

Despite its size, the sound character of the KM 56 is strikingly similar to a U 47 but with slightly less fullness in the bass and a more detailed top. The KM 54, which is cardioid only, is a brighter, slightly more aggressive-sounding mic, which works great for close-miking guitars and other acoustic instruments where you want to minimize the boominess resulting from the proximity effect when you get close. The KM 53 was the omnidirectional member of the family.

Neumann stopped making the KM 54's all-metal diaphragms in 1969, in large part because their ultra-thin construction was so fragile. Since many thousands of KM 54 microphones had been sold, all of Neumann's stock of replacement capsules was then exhausted in attempting to keep those microphones functional.

By 1970, Neumann devised an adapter to let them use the Mylar capsules of the KM 60/70/80 series on the bodies of KM 53 or 54 microphones, along with a slight wiring change to correct the polarity of the output signal. This modification prevented a KM 53 or 54 with a broken capsule from becoming entirely useless, but the resulting microphone doesn't sound like a KM 53 or KM 54, instead sounding more like the model whose capsule is being used with a more limited dynamic range.

Neumann U 87

The U 87 is probably the best known and most widely used Neumann studio microphone. First introduced in 1967, it's equipped with a large dual-diaphragm capsule with three directional patterns: omnidirectional, cardioid, and figure-8. These are selectable with a switch below the head grille. A 10dB attenuation switch is located on the rear, which enables the microphone to handle sound pressure levels up to 127dB without distortion (see Figure 2.9).

Figure 2.9: Neumann U 87
Courtesy of Neumann USA

The U 87A has lower self-noise and higher sensitivity (in other words, for the same sound pressure level, it puts out a higher output) than the original U 87. The overall sound of the two models is generally quite similar. The original U 87 could be powered by two internal photoflash batteries (22.5-V apiece). That option was removed in the U 87A model.

The latest model, the U 87 Ai model, features a a higher sensitivity, greater dynamic range and less noise, but with a slightly different frequency response in the midrange.

AKG D 12/112

Introduced in 1953, the D 12 was the first dynamic microphone with cardioid characteristics. Originally a standard choice for vocal applications for more than a decade, the mic's proximity effect and slightly scooped midrange eventually made it a favorite choice for rock kick drums (see Figure 2.10).

Figure 2.10: AKG D 12
© 2023 Bobby Owsinski

The AKG Model D 112 is a descendent of AKG's earlier D 12 dynamic microphone, widely known for its ability to handle high-level signals from bass drums and bass guitars in the studio. The microphone was designed with a low resonant frequency and the ability to handle very high transient signals with extremely low distortion. High-frequency response has been tailored to keep both bass drum and bass guitar clearly distinguishable in the mix. A built-in windscreen makes the D 112 also suitable for other high SPL applications (see Figure 2.11).

Figure 2.11: AKG D 112
Courtesy of AKG Acoustics

AKG C 12/Telefunken ELA M 250/251

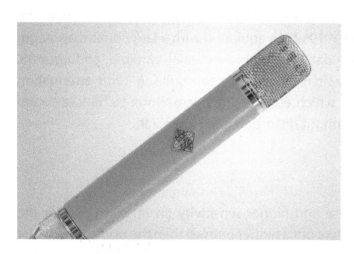

AKG, which stands for Akustische und Kino-Geräte (Acoustic and Film Equipment), developed the original C 12 condenser microphone in 1953, where it remained in production until 1963 (see Figure 2.12).

Figure 2.12: Telefunken C12
Courtesy of Telefunken

The original CK12 capsule membrane was 10-micron-thick PVC but was later changed to 9-micron-thin Mylar. The amplifier design was based around a 6072 tube. The C12 had a remotely controlled pattern selection from omni to bidirectional via the selector switch located in a box between the microphone and the power supply.

In 1965, AKG developed the C12A, which shared the capsule design with the original C 12 (but not the electronics), but had a whole new body style-one that would foreshadow what was to become the 414 series.

In 1959, Telefunken commissioned AKG to develop a large-diaphragm condenser microphone, which soon became the ELA M 250 (which stands for electroacoustic microphone). This design incorporated the same CK12 capsule but in a wider body with a thicker wire mesh grille, with a two-pattern selector switch (cardioid to omnidirectional) placed on the microphone. The ELA M251 added a third bidirectional pattern to the switching arrangement. The 251E model indicates an export model and incorporates a 6072 tube amplifier. A plain 251 indicates the use of the standard German AC 701K tube amplifier.

There were approximately 3,000 ELAs (M and M 251s) built between about 1964 and 1969, although Telefunken's original records were lost so no one knows for certain.

Because of their full-bodied yet crisp sound, the C12 and ELA M250/251 microphones have since become some of the most expensive and highly prized vintage tube mics on the market today.

AKG C451

With a styling reportedly based upon a large cigar smoked after a creative wine-tasting session, the 451 series was AKG's first FET preamplifier mic featuring interchangeable capsules. Most 451's are usually found with CK-1 cardioid capsules, although some can be found with CK-2 omni capsules, CK-9 shotgun capsules, or the CK-5, which was a shock-mounted version with a large protective windscreen/ball end for handheld use (see Figure 2.13).

Figure 2.13: AKG C451
Courtesy of AKG Acoustics.

The 452 was identical to the 451 except it had an amplifier that required 48-volt phantom power, while the 451 could run on anything from 9 to 48 volts. As 48-volt phantom power became the standard, the 452 gradually replaced the 451.

Subsequent replacement versions of the 451 are the 460 and 480 series. These both feature flatter frequency response, quieter preamps, and more headroom, but they never gained the same acceptance as the original 451. A reissue of the 451 with the popular CK-1 capsule, the C451 B, is currently being produced.

AKG 414 Series

Basically the transistor version of the C12A (see Figure 2.14), which used a miniature tube called a Nuvistor, the 414 has gone through many updates and changes through the years. Starting off as the model 412 in the early '70s, the mic was the first to use phantom power (12 to 48V DC) instead of an external power supply. This version was susceptible to radio-frequency interference if not modified, and since the grill housing was made out of plastic, was prone to cracking.

Figure 2.14: AKG C12A
Courtesy of AKG Acoustics

The C414EB (Extended Bass) was introduced in the late '70s and consisted of an all-metal silver housing. Early versions had the original brass CK-12 capsule, while the later ones had a plastic injected type. This mic was able to operate on phantom power of 9 to 48 volts. Of all 414 versions, this one seems to be the most desirable (see Figure 2.15).

Figure 2.15: AKG 414 XLS
Courtesy of AKG Acoustics.

The C414EB/P48 appeared sometime in the early '80s and is a 48V-only phantom-power version of the C 414 EB. The housing is black.

C 414 B-ULS stands for Ultra Linear Series and has been in production since the late '80s. This mic has a redesigned preamp that provides a flatter frequency response.

The C 414 B-TL is the exact same mic as the C 414 B-ULS except it uses a transformerless output stage, which gives the mic a slightly lower frequency response.

The C414 B-TL II is the same mic as the C414 B-TL except it uses a new version of the CK-12 plastic injected capsule, which was designed to give a high-end boost to emulate the sound the original brass CK-12.

The C414 B-XLS is the latest version in the 414 family, featuring a slightly larger grille and body, and decreased handling noise and higher sensitivity. It also incorporates an entirely new electronics section that does away with the old mechanical switches and replaces them with flush-mounted electronic pushbuttons for pattern, attenuation, and low-pass filter. The XLS has five pattern choices, including a new wide cardioid position, attenuation choices of 0, -6, -12, or -18dB, and low-pass filter positions of flat, -12dB at 40Hz, -12dB at 80Hz, and -6dB at 160Hz.

The C414B-X II is the same as the XLS version except for a pronounced presence peak at 3kHz.

All 414's feature a multi-pattern switch on the front and a 10dB pad and high-pass filter switch on the rear of the casing.

Sony C-37A

Introduced in 1955, the C-37A was Sony's answer to the Neumann U 47. In fact, the original C-37A was considered the finest general-purpose condenser mic available until Neumann answered it with the U 67, which incorporated many of its features. These included filtering the high-frequency resonance, the shape of the windscreen, and the built-in low-cut filters (see Figure 2.16). It takes its name from its diaphragm element, which is 37 microns thick.

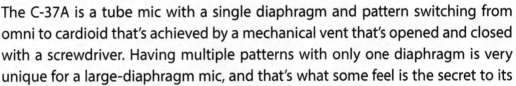

Figure 2.16: Sony C-37A
© 2023 Bobby Owsinski

The C-37A is a tube mic with a single diaphragm and pattern switching from omni to cardioid that's achieved by a mechanical vent that's opened and closed with a screwdriver. Having multiple patterns with only one diaphragm is very unique for a large-diaphragm mic, and that's what some feel is the secret to its sweet sonic character. The C-37A was first manufactured with the power supply model CP-2, which used a tube for the main high-voltage supply. This was later replaced with a completely solid-state power supply - the model CP-3B.

The C-37P was introduced in 1970 and was mechanically identical to the tube C-37A except that its preamp used an FET instead of a 6AU6 tube. This version of the mic is far less desirable than the original A model. In 1992, Sony introduced the C-800, which aside from its body style was a near-perfect clone of the C-37A. The company was not good about communicating that fact, and the mic was eventually discontinued, although the multi-pattern C-800G is still available (see Figure 2.17).

Figure 2.17: Sony C-800G
Courtesy of Sony Professional Products

Schoeps M 221B

The Schoeps M 221B is an interchangeable system in which 10 different capsules with different directional or frequency-response characteristics can be attached to a tube amplifier body. Schoeps mics are known for their sweet, smooth sound, especially off-axis, and the 221B is a particularly good example of that sound (see Figure 2.18).

Figure 2.18: Schoeps M 221B
Courtesy of Schoeps GmbH

As with so many vintage mics, the condition of the capsule membranes is very important in this series, since Schoeps no longer manufacturers the M 221 and can no longer replace the capsules. The model that replaces it, the M 222, uses the modern Colette series of capsules and has a different sound as a result.

Figure 2.19: STC/Coles 4038
Courtesy of Coles Electroacoustics

STC/Coles 4038

The 4038 ribbon microphone was designed by the BBC in 1954 and was originally manufactured by STC and most recently Coles. Long the favorite of British engineers and used on countless records in the '50s and '60s, the 4038 never found its way into many American studios. Somewhat on the fragile side, the 4038 excels on brass, percussion and as a drum overhead mic (see Figure 2.19).

Shure SM57

Over the years, the Shure SM57 has established itself as the second most popular microphone in the world (after its brother, the SM58). It is widely used in both live sound and recording applications, particularly on vocals, guitar amplifiers, and snare drums. It is used in such a large variety of situations that it often tops engineers' lists of "the one microphone to be stranded with on a desert island" (see Figure 2.20).

Figure 2.20: Shure SM57
Courtesy of Shure Incorporated

With a heritage dating back to the original Unidyne capsule used in the Shure Model 55 in 1939, the cardioid dynamic SM57 utilizes an updated Unidyne III capsule first used on the Model 545 in 1959.

Introduced in 1965, the SM57 was offered as a high-quality microphone for speech applications in broadcast, recording, and sound reinforcement. Though the microphone achieved some acceptance in the broadcast field, its ultimate success was with live sound applications and recording. By about 1968, the SM57 had been discovered by the fledgling concert sound industry.

To engineers then and now, the microphone provided a wide frequency response with an intelligibility-enhancing 6kHz presence peak (necessary for the poor sound systems during the time when it was developed), a very uniform cardioid polar pattern to minimize feedback and other unwanted pickup, and an affordable price. (The original retail price was about $85 with cable.)

The SM57 has not undergone a major change to its basic design since its introduction and still remains widely available.

Figure 2.21: Sennheiser MD 421-II
Courtesy of Sennheiser USA

Sennheiser MD 421

Go to any tracking date, and chances are you'll find an MD 421 on either the toms or a guitar amp. There have been three basic 421 models: the original 421 in gray, the newer 421 in black (which sounds pretty much the same as the gray), and the new model II, which sounds different from the first two. The cardioid 421 has a very useful roll-off switch located near the XLR connector. The response ranges from the flat M, or "music," position to the rolled-off S, or "speech," position. Over the years, the number of stops between S and M on the roll-off switch changed, with five being the most common (see Figure 2.21).

Sennheiser MD 441

The MD 441 is designed to have more upper midrange and less low-frequency response than the 421, as well as extremely directional response. When used in a live situation, the gain before feedback is indeed impressive. Because of its supercardioid pickup pattern, the 441 excels as a scratch vocal mic and is often used both on top and under a snare drum (see Figure 2.22).

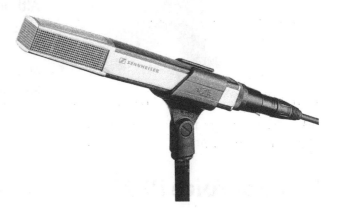

Figure 2.22: Sennheiser MD 441
Courtesy of Sennheiser USA

Sennheiser MKH 416

Originally developed in the 1970's, the MKH 416 shotgun microphone has dominated location recording and voice-overs in Hollywood and around the world (see Figure 2.23).

One of the reasons is that the mic is so good at rejecting off-axis sounds while maintaining sensitivity and detail at long distances. This relies on the 'interference tube' method where the long body is perforated with grill slots. When you point the microphone at a sound source, the sound passes down the length of the tube directly to the mic's diaphragm. The off-axis sounds hit the grill slots on the side of the microphone hitting the diaphragm at various times, causing phase cancellation.

The 416 is often chosen because of its excellent sensitivity in the 2,000 to 8,000Hz frequency range, which is optimized for capturing dialog on film or television set. It also has excellent bass performance and clarity with minimal noise, which is why the microphone has become a favorite among voice-over artists.

Although the 416 isn't used in the studio that often, you will see it on occasion as an overhead mic pointed directly over the snare drum. That said, any recording studio that does voice-over work will usually have a 416 on hand.

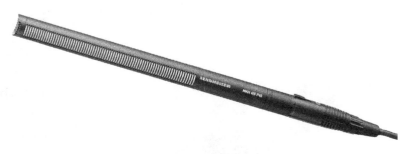

Figure 2.23: Sennheiser MKH 416
Courtesy of Sennheiser USA

Figure 2.24: Beyer M 160
Courtesy Beyer Dynamic

Beyer M 160

The Beyer M 160 is one of the so-called "modern" ribbon mics. Utilizing dual ribbons to attain a hyper-cardioid pickup pattern, the M 160 is a lot more rugged than its ribbon predecessors (you still have to be careful, though). Although used primarily on acoustic instruments by most engineers, the M 160 has nonetheless gained a sterling reputation for use on guitar amplifiers. There is also a figure-8 version of the M 160 called the M 130 (see Figure 2.24).

Electro-Voice RE20

A staple of any mic locker, the E/V RE20 is a large-diaphragm dynamic mic featuring an E/V innovation called Variable D or Variable Distance. Thanks to the ports along the sides of the microphone, Variable D allows the mic to reduce proximity effect while maintaining a flat frequency response regardless of how close to or far away from the sound source it's placed. A first choice of broadcasters since its introduction, the RE20 has found its way into the studio as a kick drum mic, vocal mic (a favorite of Stevie Wonder), floor tom mic, and many places that a condenser mic would usually be used (see Figure 2.25).

Figure 2.25: E/V RE20
© 2023 Bobby Owsinski

NEW VERSIONS OF THE CLASSICS

While every engineer agrees that the classic mics sound great (which is why some of them are still in daily use even though they may be as much as 70 years old), there are just not enough of them to go around. This has driven up the prices to the point where only the most successful artists, studios, producers, and engineers can afford them. As a result, there are many new versions of the old classics currently made by small boutique manufacturers at a considerably lower cost than the originals.

While some manufactures get closer to the original sound than others, all are sufficiently in the ballpark in a way that makes these mics a pretty safe purchase. After all, even the originals didn't sound the same from mic to mic, and if they were made exactly the same today (a very few still are, in fact), they wouldn't sound the same as the ones made 20 to 60 years ago, due to the aging of the parts inside.

Even with the original specifications, new versions of the classics sound a bit different because many of the parts in the original models just aren't made any more. Capacitors and diaphragms are made differently, transformers and inductors cannot be made the same due to the latest OSHA safety laws, and VF14 vacuum tubes (for instance) haven't been made in at least 50 years. That being said, today's boutique microphone makers do an amazing job of getting close to the original sound with the parts available on the market today.

It should be noted that the AKG C 12/ELA M 251, Neumann U 47 and U 67, and RCA 77 are the most copied, since many of the others, such as the Shure SM57, E/V RE20, Beyer M 160, and Sennheiser MD 441, are still made exactly as they were when they were first introduced. Others, such as the Sennheiser MD 421, AKG C 414, and C 451, are currently manufactured but have changed their sound over the years (some for the better, contrary to popular belief) due to continual upgrades.

Let's take a look at a few of the new classic microphone manufacturers and their offerings.

Audio Engineering Associates

Designer and founder Wes Dooley has long been both a connoisseur and an expert on ribbon microphones, and his new microphones based loosely on the RCA microphones of old (see Figure 2.26) have drawn rave reviews from recording luminaries around the world. Wes also completes the line with a couple of microphone preamps (the RPQ2 and TRP2) specially designed to complement ribbon mics. You can learn more at ribbonmics.com.

Figure 2.26: AEA R44
Courtesy of Audio Engineering Associates

Austrian Audio

The venerable AKG was purchased by the Harmon International conglomerate in 1994, but when the company closed its plant in Vienna in 2017, 22 AKG employees formed Austrian Audio. The idea was to carry on the AKG tradition without the burden of outside interference so they could get back to the company's standard of excellence.

Figure 2.27: Austrian Audio OC818
Courtesy of Austrian Audio

Since then, Austrian Audio has released a variety of transducer products including microphones and headphones. Its first product (and still its flagship) is the OC818 (see figure 2.27), a microphone designed as the successor to the AKG 414. It has met with worldwide acclaim and is worth checking out. Find out more details at austrianaudio.com.

UA Bock Series (Formerly Bock Audio and Soundelux)

Named after microphone maven David Bock, Universal Audio Bock Series currently manufactures an ELA M 251 clone known appropriately as a UA Bock 251 (see Figure 2.28), and U 67 and U 87 clones. Prior to being acquired by Universal Audio, Bock Audio Designs and Soundelux before that, made a variety of mic clones as well.

Bock's designs are unique in that there's no attempt to make their mics look like the mics they emulate, nor is there an attempt to exactly copy the inner workings at the expense of better performance. For more info on modern microphone philosophy and manufacturing, see David Bock's interview at the end of the chapter. Also check out uaudio.com/microphones.

Figure 2.28: UA Bock Audio 251
Courtesy of Universal Audio

Mojave Audio

Mojave Audio is an offshoot of Royer Labs in that all their products are designed by David Royer, but unlike Royer, none are ribbons. The company has a number of mics that don't physically resemble the classics, but they sure sound like them. The MA-200 is a large-diaphragm tube condenser that knowledgeable users say has a sound similar to the famed U 67 (see Figure 2.29), while the MA-100 is a small-diaphragm tube mic along the lines of the Schoeps M 221 or Neumann KM 56. You can check them out at mojaveaudio.com.

Figure 2.29: Mojave Audio MA-200
Courtesy of Mojave Audio

Pearlman Microphones

Pearlman Microphones makes a hand-built (as most boutique items are) version of the U 47 that comes in various flavors and prices. For more info go to pearlmanmicrophones.com (see Figure 2.30).

Figure 2.30: Pearlman TM 1
Courtesy of Pearlman Microphones

Peluso Microphone Lab

Figure 2.31: Peluso Microphone Lab CEMC-6
Courtesy of Peluso Microphone Lab

Peluso Microphone Lab is the brainchild of John Peluso, who, like most other boutique microphone designers, is a long-time mic repairman. Like David Bock, John makes clones of not only the popular 251s and U 47s, but the RCA 77 and the Schoeps CMC line as well. Peluso mics differ from those of other boutique manufacturers in that they're fairly inexpensive as well. For more information, go to pelusomicrophonelab.com (see Figure 2.31).

Telefunken Elektroakustik

Telefunken Elektroakustik is the modern incarnation of the original Telefunken division that originally distributed the classic highly-sought ELA M 251 and U 47. Today the company builds extremely faithful reproductions of those mics (some at prices similar to their vintage forbearers) in several versions and more. For more information, go to telefunken-elektroakustik.com (see Figure 2.32).

Figure 2.32: Telefunken Elektroakustik U-47
Courtesy of Telefunken Elektroakustik

Wunder Audio

Wunder Audio makes not only microphones based on the classics, but microphone preamps, EQs, and even a large-format analog console as well. The microphone line consists of the standard U 67, 251, and U 47 clones, but also the revered M 49/M 50 and U 47 FET. For more information, go to wunderaudio.com (see Figure 2.33).

Figure 2.33: Wunder Audio CM49
Courtesy of Wunder Audio

Warm Audio

Although Warm Audio has been making affordable pro audio gear since 2011, the company didn't release its first microphone until 2016. Since then it's gone on to make a version of just about every vintage mic you can think of, and at very affordable prices as well. The company was one of the first to release a clone of the Sony C-800G (called the WA-8000 - see Figure 2.34), a favorite of hip hop and R&B vocalists everywhere.

Figure 2.34: Warm Audio WA-8000
Courtesy of Warm Audio

There are many other manufacturers of either direct clones or vintage sounding microphones. As always, you should always listen before you purchase, since the sound from mic to mic may differ even on the same model.

MODELING MICROPHONES

We'd all love a mic locker full of choice vintage classics, but that's an impractical wish, both from the standpoint of trying to find one for sale, paying the high price it demands, then hoping that the mic you purchased is in good working condition. It's getting more and more difficult to find a vintage mic that doesn't show the effects of 50 plus years of age, and harder to restore them to their former glory as well.

That's why the new class of modeling microphones are so interesting. Not only do they provide a sound close to the real thing, but they even let you change the sound later after recording.

Modeling microphones are best thought of as a system since they consist of both the microphone and the software that allows you to dial in the particular mic type you're looking for. Even better is the fact that in some cases you're also able to record on several different tracks at once with different mic models, so it's a snap to AB later. Right now there are three modeling microphone systems, but no doubt there will be more to come in the future.

Townsend Labs Sphere L22

Debuting in 2016, the first modeling mic was the Sphere L22 by Townsend Labs (see Figure 2.35), which has since been acquired by Universal Audio. The mic itself is very well built, with two large gold-sputtered diaphragms for stereo recording. The two capsules also provide the option of setting both sides to a different mic model, each with its own set of controls in the plugin. The associated software plugin allows the mic to emulate 34 highly desirable dynamic, ribbon and condenser mics, including the Neumann U47, U67 and U49, Shure SM57 and SM7A, Sony C800G and C-37A, AKG C12, 414 and 451, Electro-Voice RE20 and Sennheiser MKH 416, just to name a few. Expansion packs include UA's Ocean Way Microphone Collection plugin with 12 additional mic models, and the Bill Putnam Microphone Collection with nine additional mic models.

Figure 2.35: Universal Audio Sphere L22
Courtesy of Universal Audio

Slate Digital VMS

The Slate Digital Virtual Microphone System (VMS) consists of either a large (ML-1) or small (ML-2) diaphragm microphone component along with their Virtual Mix Rack (VMR) plugin, which provides the ability to choose the various mic models. Each mic has a different set of emulations available,

with the VMS ML-1 including the U47, C800G, SM7B and C12, among other popular models, and the VMS ML-2 including the SM57, SM7B, RE20 and various AKG's. The VMR is also a powerful modular mix processing and effects platform as well.

Figure 2.36: Slate Digital VMS
Courtesy of Slate Digital

Antelope Audio Edge

Antelope's Edge series of modeling microphones consist a basic large diaphragm condenser mic in various configurations. The Edge Solo is a single capsule cardioid mic, while the Duo and Quadro have two capsules that allow for multi-channel recording or stereo. The Edge Go has onboard DSP that provides real-time EQs, channel strips, and effects, and also connects via USB-C, eliminating the need for a preamp. 18 microphone models are available for each microphone.

Figure 2.37: Antelope Audio Edge Quadro
Courtesy of Antelope Audio

Modeling Software

There are also plugins now available that will emulate the classic microphones while using a normal mic you own instead of one from the manufacturer. These include IK Multimedia's Mic Room, Antares Mic Mod, Waves The Kings Microphones, and others. At least at this time, these emulations tend to only get you in the ballpark sound-wise, and don't sound nearly as close to the real thing as the systems that include a dedicated microphone.

Figure 2.38: IK Multimedia MIc Room
Courtesy of IK Multimedia

The sound and versatility that modeling microphones provide open up new sonic possibilities for users with a limited budget, along with unprecedented versatility. We will undoubtedly see more entrants into the market as well as more occupying mic cabinets in the future.

43 | Common Microphones

THE NEW CLASSICS

While many believe that microphone technology hasn't really improved in at least 30 years, a host of new mics have taken the technology to the next step. Some have even become classics in their own right and can be found as standard equipment in mic lockers the world over.

Audio-Technica AT4050/4033

The AT4050 is a large-diaphragm multi-purpose, multi-pattern condenser mic that's found its way into mic lockers everywhere (see Figure 2.39). It's open and airy top end, low noise, and ability to take punishing SPL levels have made this a go-to mic when your usual favorite just isn't cutting it. It's also relatively inexpensive compared to the similarly featured German and Austrian favorites. The cardioid-only 4033 is a less expensive version of the same mic.

Figure 2.39: Audio-Technica AT4050
Courtesy of Audio-Technica

Heil Sound PR 40

Another newer mic that has caught on for kick drum use is the Heil Sound PR 40 (see Figure 2.40). The PR 40 incorporates a large 1-1/8th-inch dynamic element for an extended low-frequency response as well as a presence bump from 2.5k to 4.5kHz. It's also capable of handling very high SPL levels, and its super-cardioid pattern provides excellent back-side rejection. Many feel that the response of the PR 40 has sort of a "pre-EQ" built into it that makes EQing later either unnecessary or a lot more gentle than with other mics.

Figure 2.40: Heil PR 40
Courtesy of Heil Sound

Royer R-121

Introduced in 1996, the R-121 is the first radically redesigned ribbon microphone in that it has a higher output than older ribbons, is a lot more rugged, and can take all the SPL you can hand it. You'll see it used where the old favorite ribbons are used (overheads, brass), but in some new places too, such as kick drum and guitar amps (see Figure 2.41).

Figure 2.41: Royer R-121
Courtesy of Royer Labs

Shure Beta 52A

The Beta 52A (or B52, as some call it) is the first mic to give the revered AKG D 112 some competition as a kick mic. The mic is specially designed for kick and bass with an EQ curve built in to attenuate the 300 to 600Hz "boxy" frequencies and boost around 4 kHz for presence. It can also handle extremely high SPL levels up to 178dB (see Figure 2.42).

Figure 2.42: Shure Beta 52A
Courtesy of Shure

Shure SM81

Although not truly a recent mic (it was introduced in 1978), the small-diaphragm SM81 condenser has been slowly building favor through the years, until it's now often used whenever a small-diaphragm mic is called for. Known for its flat frequency response from 20Hz to 20kHz, low noise and RF susceptibility, the SM81 is ruggedly constructed and operates over a wide variety of temperatures. It has a built-in 10dB pad and a switchable flat 6 or 18dB per octave high-pass filter (see Figure 2.43).

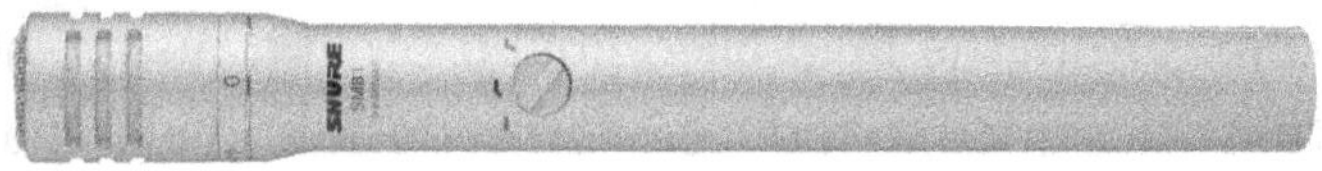

Figure 2.43: Shure SM81
Courtesy of Shure

Figure 2.44: Yamaha SKRM-100 Subkick
© 2023 Bobby Owsinski

Yamaha SKRM-100 Subkick

The Yamaha SKRM-100 was created as an answer to the subkick phenomenon that started due to the burning desire to get more bottom end from the kick drum without having to crank up the EQ. The unit was originally designed to capture the frequencies below 60Hz, which is something that you feel more than you hear.

The trend started when a few engineers began to take the woofer from a Yamaha NS-10M monitor, use the magnet to attach it to a mic or drum stand about 2 inches in front of the bass drum, and plug it into a direct box. This wasn't a new idea by any means, as engineer Geoff Emerick tried something similar on Beatles records ("Rain" and "Paperback Writer") in the '60s. The problem is that Yamaha no longer produces the NS-10, and the factory that made the woofer has closed. Engineer Russ Miller took the idea to Yamaha, who manufactured a unit that contains a 6-1/2-inch speaker mounted inside a 7-ply maple shell with black mesh heads, so it's actually a speaker mounted inside a drum shell (see Figure 2.44).

Unlike a homemade subkick (see Figure 2.45), the Yamaha subkick is tunable, but some engineers argue that the sound still wasn't as good as what you can get from just a raw NS-10 woofer.

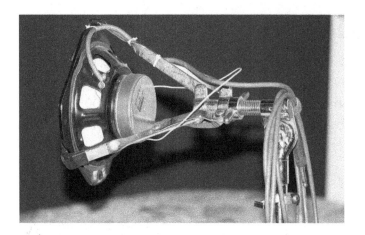

Yamaha has since discontinued the SKRM-100 thanks to the fact that there are so many inexpensive sub-bass plugins that can do the same job with less hassle. You still run into some form of the subkick every now and then (the Avantone Pro Kick and the DW The Moon are different versions of the same idea), so it's worth knowing about.

Figure 2.45: A homemade subkick
© 2023 Bobby Owsinski

Sony C800G

Released in 1992, the tube-based Sony C800G took some time to be discovered, but once it was it soon became the go-to mic for rap, pop and R&B thanks to its bright, detailed sound. It's unusual look comes from the fluid-filled finned cooling system at the mic's rear that's designed to reduce internal noise and eliminate distortion. It has two pickup patterns - cardioid and omnidirectional. This is one of the last high-end microphones to be cloned.

TIP: The finned cooling system of the C800G is filled with fluid, which travels to a different position when the mic is turned on its side or placed upside down. This will cause the frequency response of the mic to change.

Figure 2.46: Sony Audio C800G
© 2023 Bobby Owsinski

INEXPENSIVE MICROPHONES

One of the more interesting recent developments in microphones is the availability of some extremely inexpensive condenser and ribbon microphones in the less than $500 category (in some cases even less than $100). While you'll never confuse these with a vintage U 47 or C 12, they do sometimes provide an astonishing level of performance at a price point that we could only dream about a few short years ago. That said, there are some things to be aware of before you make that purchase.

Quality Control's The Thing

Mics in this category have something in common: Either they're entirely made or all their parts are made in China, and to some degree, mostly in the same factory. Some are made to the specifications of the importer (and therefore cost more), and some are just plain off-the-shelf. Regardless of how they're made and to what specs, the biggest issue from that point is how much quality control (or QC, also sometimes known as quality assurance) is involved before the product finds its way into your studio.

Some mics are completely manufactured at the factory and receive a quick QC just to make sure they're working, and these are the least expensive mics available. Others receive another level of QC to get them within a rather wide quality tolerance level, so they cost a little more. Others are QC'd locally by the distributor, with only the best ones offered for sale, and these cost still more. Finally, some mics have only their parts manufactured in China, with final assembly and QC done locally, and of course these have the highest price in the category.

You Can Never Be Sure Of The Sound

One byproduct of the rather loose tolerances due to the different levels of QC is the fact that the sound can vary greatly between mics of the same model and manufacturer. The more QC (and the higher the resulting price), the less difference you'll find, but you still might have to go through a number of them to find one with some magic. This doesn't happen with the more traditional name brands that cost a lot more, but what you're buying (besides better components in most cases) is a high assurance that your mic is going to sound as good as any other of the same model from that manufacturer. In other words, the differences between mics are generally a lot less as the price rises.

The Weakness

Two points contribute to what a mic sounds like: the capsule and the electronics. The tighter the tolerances and the better the QC on the capsule, the better the mic will sound and the closer each mic will sound to another of the same model.

The electronics is another point entirely in that a bad design can cause distortion at high SPL levels and limit the frequency response, or simply change the sound enough to make it less than desirable. The component tolerances these days are a lot tighter than in the past, so that doesn't enter into the equation as much when it comes to having a bearing on the sound. In some cases, you can have what could be an inexpensive, great sounding mic that's limited by poorly designed electronics. You can find articles all over the Web on how to modify many of these mics, with some techniques that make more of a difference to the final sound than others. If you choose to try doing a mod on a mic yourself, be sure that your soldering chops are really good, since there's generally so little space that a small mistake can render your mic useless.

Some Good Choices

While new inexpensive mics are coming on the market every day, here are a few that users have been generally pleased with.

Behringer B-1

While everything that Behringer makes is in the budget category, the large-diaphragm B-1 condenser seems to be an item that's found some acceptance. It's a remarkable value in that you get a carrying case, shock mount, and windscreen along with the mic.

Cascade FAT HEAD

While other budget-mic companies have concentrated on condenser mics, Cascade Microphones have gone in a different direction with their inexpensive ribbon mics. The FAT HEAD is just about the least expensive ribbon mic on the planet (complete with a wooden box) but still has a host of very satisfied users.

MXL MCA SP1

Marshall Electronics were one of the first to bring out what looks to be a large-diaphragm condenser mic at an ultra-inexpensive price (it's actually a small diaphragm in a large body). The MCA SP1 is one of the best deals you'll ever find for a mic like this, and there are many mods available to improve the sound quality.

Oktava MK-012

A small-diaphragm condenser with interchangeable capsules for different patterns, the MK-012 has become a favorite for cymbals, acoustic guitars, and choirs.

RØDE NT1-A

RØDE actually started as a Swedish company before finally settling in Australia, so it really doesn't fit into this category since all the parts except the capsules are manufactured there. That being said, they produce quality microphones for people on a budget. The NT1-A is the much improved, quieter successor to the original NT1.

sE Electronics sE2200a

Built by hand in their own factory in Shanghai, sE Electronics mics are one product in this class that's finding its way into the hands of everyday pros and name recording artists.

Studio Projects C1

The Studio Projects C series of large- and small-diaphragm condenser mics has become one that has a very high user satisfaction rating, with the C1 being the first introduced. A single 3-way switch on the rear of the microphone body enables either a high-pass filter or -10dB pad.

Audio Technica AT2020

With the AT2020, Audio Technica has managed to create one of the best of the inexpensive microphones. Made of metal instead of plastic, the AT2020 was designed for the project studio, specifically for vocals. It works well in a variety of other applications as well, and unlike other inexpensive mics, can handle high transients without distorting.

Warm Audio WA-47jr

The WA-47jr is a large diaphragm FET condenser mic with features that go beyond most inexpensive microphones. It has three polar patterns, a 10dB pad, and 70Hz Hi-Pass filter, and even comes with a basket-style isolation mount more commonly found with much more expensive microphones.

CLEANING AND DISINFECTING YOUR MICROPHONES

Thanks to the pandemic, we're now more aware than ever that our microphones need to be as germ-free as possible. Microphone manufacturer DPA has issued some guidelines for disinfecting your microphones and accessories that I've paraphrased and excerpted here. Let's dig in.

Cleaning versus Disinfecting

First understand the differences between cleaning and disinfecting. Cleaning means to remove dirt, while disinfecting means killing germs (we want both). Clean first, then disinfect.

DPA suggests that the best way to clean the mic is to wipe down its surfaces with soap and water, which also destroys germs in the process. Take care not to get any on the capsule. Let dry for 72 hours.

Disinfecting

This is basically the same as cleaning except for the solution used. In this case, wipe down the microphone with a cloth moistened with 70% isopropyl alcohol. *Be sure you don't get any on the mic's capsule.* If you go to the drug store you'll find 99% alcohol and you might think it will work better because it's stronger. That's not the case though, since it evaporates too quickly to kill any germs. A diluted solution will actually work much better. You can still use the 99% isopropyl alcohol by adding 20% water to it, which will allow it to linger long enough to do its job.

While DPA suggest 80% alcohol, the U.S. Center for Disease Control says that 70% (which is more readily available in stores) is effective in killing germs like the coronavirus.

Wind Screens

Foam wind screens should be cleaned by gently removing them from the microphone, then cleaning with soap and water. Leave them to dry for 72 hours, which is long enough for any remaining germs to die. Unfortunately you can't use isopropyl alcohol here as it will damage the foam.

Cables

It's easy to overlook cleaning cables but they're things that we touch all the time without even thinking about it. Like foam wind screens, you can't use alcohol to clean them because it will break down the cable jacket and eventually make them brittle. DPA suggests using olive or coconut oil, but good old fashioned soap and water works as well. Like everything else not exposed to isopropyl alcohol, let dry for 72 hours to be sure they're germ-free.

DPA also suggests that you can speed up the 72 hour waiting process by placing the gear in an oven set at 140 degrees for an hour. The problem is that microphone capsules might age slightly from doing that, but if you're in a hurry it may be worth it.

Cleaning and disinfecting your microphones is probably something that you never thought about, but it's always a good practice to do so regularly. It's not difficult and just takes a little time, and it will go a long way to keeping you and your clients healthy.

MEET MICROPHONE DESIGNER DAVID BOCK

After maintenance stints at such prestigious facilities as the Hit Factory in New York City and Ocean Way in Hollywood, Bock Audio Designs (formerly Soundelux and now part of the Universal Audio family) founder and managing director David Bock went from repairing vintage microphones to manufacturing them. David has utilized his expertise to produce updated versions of the studio classics like the ELA M 251, U 47, U 47 FET, and U 67. David was kind enough to share some of his insights to the inner workings and differences between classic microphones and their modern counterparts.

How did you get involved in designing your first mic?

I had been modifying some Chinese microphones that someone had brought into the country about 15 years ago, and some people got excited about the results. I had specialized in repairing microphones up until that point, so modifying them was the next step. Saying to myself, "Why are we always using the same single-triode design for microphones? Let's do something different," was the final step to building them.

What actually makes a vintage microphone so special?

There are a couple of things that go into that. The bottom line is that the '50s were really the golden age of audio design. Those guys really did know what they were doing when they designed a lot of the key gear that people are still using. They used a lot of the correct techniques, and they had the luxury of decent materials and the time to research things properly.

There is a tone to these things that is harder and harder to duplicate. Not impossible, just harder and harder. They had tubes back then that are harder to get now. The available selection of materials was a lot greater back then. Then there's the element of chance. Why would someone pay $20,000 for a 251? Well, maybe that particular 251 really does sound unique because AKG's production was so sloppy and the capsules were so poorly machined that you're bound to get one that excels beyond everything else and the rest are just kind of average. Now we have CNC machines that can make these tiny little holes on the capsule backplate all the same, which AKG really couldn't do at the time.

When you set out to build a mic, how do you determine what you're going to copy?

When I started building microphones, there was no copying intended, it was merely to forge new ground. Everything was defined by economic and production parameters as well as a little ignorance, since I hadn't been in the manufacturing game that long.

That's what I was able to do initially but that wasn't my goal, so the first few microphones [the U95 and U99] established the company enough so I could get to that goal. Once I was able to get there, it was time to emulate a few of the classics that everyone used on great records. I had a client who was a 251 freak that kept bugging me to build one, and it became a several-year obsession for both of us. That's what led to that first copy, but it was not a short process.

In the case of the 47, which came after the 251, it was even a longer process. There were a lot of things that had to come together since it's such a complicated construction with a lot of parts. In some ways it was a little easier though, since I had repaired so many of them, and as a result I had a better sense for what sounded good or bad.

As you were trying to build an updated version of a vintage microphone, were you trying to copy everything including the circuitry and trying to get that as close to the original as possible, or were you trying to just make it sound like the original without worrying how you got there?

The sound comes first, but that's not the whole story. The first thing I had to do was try to find what makes the microphone sound the way it does. There were at least 15 points that you have to look at, it turns out, if you're going to emulate the sound of a microphone. The first large problem is, "I want to copy the sound of a 251." Well, which 251? I rented about ten 251s here in town [Hollywood], and you know what? There's no such thing as a common 251. They're all totally different. I could hear it and I could measure it.

Among some of them there is a common thread, though. Frequency response is the primary guidepost because all microphones have their own signature, but frequency-response curves don't always tell you everything. You have to take frequency-response measurements not only far-field but also near-field, which strangely are not published and are completely critical to what we believe a microphone sounds like in the directional world. If you saw a proximity graph and a one-meter graph, you'd have a much better idea of what the microphone sounds like.

The dissection process continued through a lot of substitutions. You might take a power supply and substitute a different circuit topology and see what it changes, for instance. There are also a lot of measurements that you have to do. Our ability to test things today is definitely better than back when the classics were built, but it's not completely conclusive and opens up a can of worms that says, "If I can't measure it then I can't hear it," which I completely disagree with. If you worked only towards measurements, you end up with something that actually doesn't sound particularly good compared to things that were designed with listening in mind.

Finally, there are listening tests. My primary listening test is to make a recording of a drum set in a large room. I've got a couple of key locations where I place the microphone to give me an idea about the close and distant pickup characteristics. That's where you start hearing the differences. Microphone capsules are related to drums. If you took ten DW kits and you tuned them all the same, they'd still all sound different. There's a parallel you could draw towards microphones. You could tune all the snare drums and toms the same and even use measurement devices to be sure that they're the same, and yet the trained ear of an engineer can pick out the differences between them. We can lock onto things that are different about each one.

What was the hardest thing to get right?

Always the capsule, because it's so small and if you make a tiny change it makes a huge difference, but that's not to say that the capsule is 99 percent of the sound. An 87 and a 67 don't sound that similar, yet they use the same capsule.

How do you deal with parts that are no longer made?

In some cases you can replicate them. In some cases you can improve them. In some cases you have to bite the bullet and say, "I just can't get that part, so I'll have to come up with the closest thing I can." Take, for instance, something like transformer laminations. We don't have the exact laminations that they used in the original 47 transformer, but we came up with something that's a lot closer than an off-the-shelf Jensen [transformer].

Then again, you're not going so much for the part but for the effect of the part, but in some cases we've found through the substitution method that some things just have to be duplicated. Like, if you mounted the tube to a printed circuit board, it would have a different resonance than if you mounted it with two rubber mounts.

In other cases a substitution can be as good or even better. For example, in our 251 we use a large core output transformer with the same turns ratio as the small transformer used in the original. That gives us a little less distortion and a lot more headroom in the low end. At the risk of not being historically accurate but being a lot more useful for today's recordings, I made a decision saying "I'd rather have the headroom" because it didn't affect anything else.

The original 251s were made out of plastic that could disintegrate in your hands. That's not acceptable, so ours are metal-framed. And the way we power the heater on the tube is different than the way they did it in 1960, but we get 6 to 10dB less noise overall because of it, so that's a useful improvement. So we try to maintain faithfulness to the vintage sound, and we'll make an improvement where we can.

What's the biggest difference in the way microphones are made today from the way the classics were made?

Mass production and availability of quality materials. Also, the need for profitability on a corporate level seems to affect how things are made a lot. I've seen the way Neumann microphones are built, and they're very different from the way they used to be. Because of the way they built their microphones in the '50s and early '60s, [it] will allow me to be able to keep those microphones running for a long time. Not so with the newer microphones. They still make a great capsule, but they don't make the microphone the same in terms of construction. They're built for ease of production and lowest cost. It's true almost across the board.

So if we were to make a broad statement, microphones are not made as well today as they were 50 years ago.

No, they're not. Even if you had a "cost is no object" attitude, you still don't even have the same metals available. The quality of brass is different now from what they used in the '50s and '60s, for instance, and an equivalent can't be found.

What is the most critical part of a microphone?

The capsule is the most critical, but electronics play a big part. You can have a great capsule but crummy electronics, and the microphone will sound mediocre. If you have really great electronics but a crummy capsule, then you still won't have a good-sounding microphone. If you have a great capsule and great electronics, then you'll have a really good microphone at that point.

If you were to look into the future of microphone development, where would you like to see it go?

Unless someone comes up with a true digital transducer that's usable, I don't know how much more it can get refined.

What seems like an improvement sometimes doesn't work at all. There have been so-called improvements along the way that were commercial flops. I'll give you a quick example. There was a microphone that AKG made, the lowly 414, which has descended into the depths of hell at this point. In their 14 revisions of this microphone, they made one called the P48 EB that used a transistorized cascode circuit. It's the most correct and stable circuit that you can use from a textbook standpoint. It's the only time that I've ever seen it used outside of a secret internal Neumann document from the early '60s, yet it's their most hated microphone of all the 414 versions. So in terms of serious evolution, I'd like to see some, but I'm a little worried the marketplace can't handle real useful advances.

Could it be that the amplifier circuit was exposing the faults of the capsule?

Possibly, since by that time they had migrated to the molded capsule that is generally accepted to be a disaster.

With the way the business seems to be going with less and less emphasis on sonic quality, will there be enough people left to appreciate what you're doing?

Anybody who is serious about the profession either evolves to a point where they say "I can use an SM57 for every track to make a record" or "I'd rather use a high-quality microphone to make a record." You're going to go one way or the other, and most people, if they stay in the business long enough, will usually gravitate to the more exclusive side.

3
BASIC RECORDING EQUIPMENT

There's nothing more important than a clean signal path when it comes to making a great recording, but what exactly does that mean? A signal path is the route that the audio signal takes from the sound you're trying to capture to where it's finally recorded. Usually that's at least a microphone, a microphone preamplifier, and a recorder (see Figure 3.1), but there are many variations on the theme; for example, if you're using a console or a DAW, there are potentially more signal processors inserted in the path (see Figure 3.2).

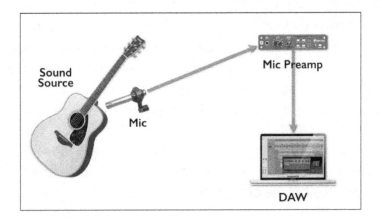

Figure 3.1: A simple signal path
© 2023 Bobby Owsinski (Source: iStock Photo, Avid, Great River Electronics)

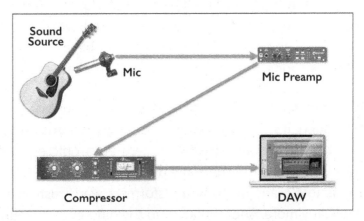

Figure 3.2: A more complex signal path
© 2023 Bobby Owsinski (Source: iStock Photo, Avid, Great River Electronics, Universal Audio)

Let's take a look at the various components that can be found in a typical recording signal path that come after the microphone.

THE MICROPHONE PREAMPLIFIER

Almost as important as the microphone is the microphone preamplifier, or mic pre, mic amp, preamp, or just simply pre. This circuit boosts the extremely small output voltage from the microphone up to a level (called line level) that can be easily sent around the studio to consoles, compressors, EQs, and DAWs.

Nearly every console and most DAW audio interfaces have mic preamps built into them, but in most cases the quality of this circuit isn't nearly as high or as costly as what's available as an outboard piece. Also, each mic pre has its own sound, and most engineers will select the mic pre and mic combination as a different color to fit the instrument and music.

Why a Separate Mic Amp?

You might ask, If nearly every DAW interface and console has its own mic pres, then why use an outboard one? There's usually only one answer: *because a dedicated unit sounds a lot better*.

An outboard mic pre generally provides higher highs and lower lows (meaning it has better frequency response), and it's sound output is clearer and cleaner. This sound-quality improvement does come at a price. While the parts of a typical mic amp in a console hover somewhere around $20 a channel (if that), an outboard mic pre can cost anywhere from $100 to several thousand dollars per channel. With the increased cost usually comes a superior design with better-quality components, as well as a larger box to put them in (generally at least 1U high with a standard 19-inch rack mount).

As with microphones, some mic pres are solid state, while some use a tube for their amplification, with both methods capable of doing the job well but ultimately sounding different because of the inherent sound of the components as well as the design that each demands.

Vintage Mic Preamps

They just don't make them like they used to. At least that's what a lot of engineers think when selecting a mic pre. There's a sound to these units that's been difficult to duplicate in modern gear, except in rare exceptions. Because of this philosophy, some of the most desirable mic amps were made in the '60s and are actually cannibalized sections of recording consoles from that era.

So why does the old stuff sound different (let's not use "better" because it's such a relative term) than the new? Very broadly speaking, it's the iron inside, meaning the transformers and inductors used routinely in older gear that gets substituted with modern electronic equivalents because of size, weight, and cost. So why not make transformers like they did back then? Although some companies try, the fact of the matter is that many of these transformers were custom-made for the particular unit and just aren't available any more. Another factor in the difference of sound can be attributed to the fact that the older units used discrete (individual) electronic components that could be properly matched to the circuit, while modern units tend to utilize cookie cutter–type integrated circuits (a complete circuit on a chip) to attempt to achieve the same end.

Here are some examples of vintage outboard mic preamps that are generally held in high esteem for their sonic qualities.

Neve 1073/1081

Of all the Neve console modules (and there are many), the Neve 1073 is probably the most famous. This unit is far more than just a mic preamp, as it's actually a channel strip featuring both a line input and an equalizer that was pulled from a console and reconfigured for outboard use. The 1073 has a 3-band equalizer with fixed EQ points and a high-pass filter. Another Neve module used often is the 1081, which differs from the 1073 in that it has a 4-band equalizer with two midrange bands and more frequency choices. Through the years, Neve made a lot of variations on the above theme as most of their consoles were custom orders, but they all had the same distinctive Neve sound (see Figure 3.3)

Figure 3.3: Neve 1073 module
Courtesy of AMS Neve Ltd.

API 312/512

API preamps (circa 1970) are classics, and most engineers will chose to use them if they are available (especially on drums). They have a tone that simply cannot be duplicated by anything else, vintage or modern, with a fat low end (due to the distortion in the old transformers) and a clear, slightly hyped high end. Although the more modern 512 sounds very similar, the older 312's are slightly fatter and smoother sounding (see Figure 3.4).

Figure 3.4: API Channel Strip
Courtesy of API

Figure 3.5:
Telefunken V72
© 2023
Bobby Owsinski

Telefunken V72/V76

Consoles of the early '60s were vacuum tube–based, and German Broadcast set a standard for preamp modules used in their consoles that was copied and used all over Europe, most notably by EMI Records in England. The Telefunken V72, V72A, V76, and V78 are the most widely used and best-loved mic pres from that period. The V72 is a dual-tube unit employing two Telefunken EF804S tubes, while the V72A used one E180F and one 5654 tube and had a bit more gain and output. The impossible-to-find V72S preamps were found inside the famous EMI REDD. 17 and 37 Abbey Road consoles that were used on Beatles' recordings up through *The White Album*. The V76/78 employs four of the EF804S tubes and has the most gain of the series (see Figure 3.5).

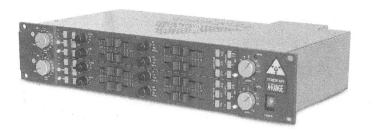

Figure 3.6: Trident A Range Reissue
Courtesy of Trident Audio Developments

Trident A-Range And 80B

Trident consoles have been the sound of rock for a long time, starting as custom-built units for the famous Trident Studios in London. The A-Range especially has a very distinctive sound, but since only 13 consoles where ever built, you never see any original mic pre modules around. The 80B was a newer less expensive console and there were a lot of them made, so you occasionally see a few original modules popping up. There's been a great demand for these modules however, so the current distributor for Trident now makes them based on the original consoles.

Modern Mic Preamps

There are many fine modern equivalents to these vintage mic preamps (some by the original manufacturers), but again, each has its own special flavor that must be chosen to suit the microphone, instrument, and music. There are basically two categories of modern mic preamp: the ones that try to emulate the unique sonic character of vintage mic pres, and the ones that try to provide the cleanest amplification without adding any character (meaning distortion). Some highly thought-of modern brands include the following.

Avalon Designs

Avalon doesn't try to sound vintage and as a result the company has built a reputation based on its own unique sound. It has several mic pres available, from the V5, M5 and AD22, to the VT-737SP, a channel strip based around vacuum tubes that has become the standard for hip-hop vocals.

Figure 3.7: Avalon VT-737SP
Courtesy of Avalon Designs

BAE

BAE builds close recreations of vintage Neve modules that sound and feel authentic. The company's 1073 module also includes the same style 3 band equalizer as the original.

Figure 3.8: BAE 1073
Courtesy of BAE

Daking

Daking manufactures the 52270B mic-pre/4-band equalizer, which differs from the other units mentioned in that it emulates the mic amp and equalizer of the famous and extremely rare Trident A Range console (see Figure 3.9).

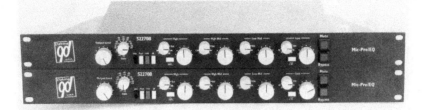

Figure 3.9: Daking 52270B Mic-Pre/EQ
Courtesy of Daking Audio

GML

George Massenburg is one of the pioneers of the audio industry (he designed the first parametric equalizer) and his GML gear is known for its super high quality and audio transparency. His 8302 is a no-frills dual channel preamp while his 2032 is a single channel unit featuring his famous parametric EQ.

Figure 3.10: GML 8302
Courtesy of George Massenburg Labs

Grace

This boutique audio company has built a reputation for building ultra-reliable products that are extremely transparent. Now, with a full product line that has grown out of their mic preamp models, anything from Grace can truly be considered "high fidelity" (see Figure 3.11).

Figure 3.11: Grace m201 mk2
Courtesy of Grace Audio Design

Great River

Although Great River makes mic preamps in the "clean modern" category as well, they also make the MP-2NV, which emulates the classic circuitry and vintage sound of the Neve 1073 module. The circuitry allows for both transformer saturation and the soft distortion resulting from pushing the input level of the unit, just like the real thing (see Figure 3.12).

Figure 3.12: Great River ME-1NV
Courtesy of Great River Electronics

Hardy

The John Hardy Company manufactures and distributes world-class microphone preamplifiers based on simple and elegant designs. Their most popular product, the M-1 mic preamp, has been impressing artists, engineers, and listeners around the world since 1987. Utilizing what many call the world's best input transformer (the Jensen JT-16-B), best op-amp (the 990 discrete op-amp), and the elimination of all capacitors from the signal path, these all combine to provide the M-1's high performance (see Figure 3.13).

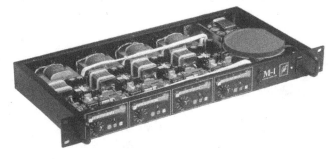

Figure 3.13: Hardy M-1
Courtesy of John Hardy Company

Millennia Audio

The HV-3C is an extremely wide dynamic range stereo microphone preamplifier intended for demanding acoustic work. With more than 12,000 channels now in use, the HV-3 is a world standard for classical and critical acoustic music recording (see Figure 3.14).

Figure 3.14: Millennia Media HV-3C
Courtesy of Millennia Media

Rupert Neve Designs

While Rupert Neve sold his original company back in 1973 (the company has changed hands several times since then), he established the new Rupert Neve Designs in 2005. His intention was not to exactly recreate what he created before but to modernize those designs to be more in-line with how engineers work today. Since then, his model 5024, 5211 and 5025 Shelford microphone preamps are highly prized.

Figure 3.15: RND 5025 Dual Shelford
Courtesy of Rupert Neve Designs

Universal Audio

Based on the legendary Universal Audio 610 modular vacuum tube console, the Universal Audio 2-610 is a new version of an old classic. The original model 610 was used on a host of '50s and '60s chartbusters, including hits by the Beach Boys and Frank Sinatra (see Figure 3.16).

Figure 3.16: Universal Audio UA 2-610
Courtesy of Universal Audio

Vintech

Vintech makes a number of units based on Neve classic designs: the X73 based on the Neve 1073 module and the X81, which is based on the 1081 (see Figure 3.17).

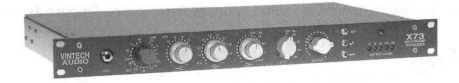

Figure 3.17: Vintech X73
Courtesy of Vintech Audio

500 Series Racks

Back in the '70s, Automated Processes Inc. (API) built consoles that were modular, and these modules became so popular that even a few other console manufacturers used them. Buy the time the '80s rolled around, a number of engineers who loved the API sound got the idea to build their own small racks in order to take advantage of these great sounding modules. At first, none of these racks were compatible, as the internal wiring of the modules was different, but soon a standard was created to make interchangeability easy. Hence, the 500 series rack was born (see Figure 3.13).

Today a wide variety of manufacturers make 500 series modules, and they have become a popular way to not only save money, but to have a variety of different types of gear available in a single portable box. 500 series racks now available with anywhere from 2 to 10 spaces, and in almost all cases the quality of the modules available surpasses that of most inexpensive analog audio gear.

That said, some gear does perform better in a dedicated rack mount form (like Figure 3.18) rather than a 500 series module. The reason is that a dedicated unit can have a power supply specifically tailored to its performance, rather than using the standard 500 series voltages. In most cases though, the performance is nearly identical and the difference isn't noticeable.

500 series modules consist predominantly of microphone preamps and dynamics, although there are some dedicated filters, analog delays, distribution amplifiers, phase alignment and other more exotic modules now on the market. Remember, virtually all of them are analog, so trying to repeat the settings means that you have to either write them down or take a picture (just like the old days).

Figure 3:18: A 6 space 500 series rack filled with RND modules
Courtesy of Rupert Neve Designs

Mic Preamp Setup

Mic preamps do only one job: amplify. Therefore, they don't usually have many controls, although the more expensive, exotic models might have extra features. The two items that every pre have in common are a *Gain* control (sometimes called *Trim*) and some type of overload indicator. Other controls that you might see are *Output Gain, Impedance, Input Pad, Phase, Hi-Pass Filter,* and more extensive metering.

Primary Controls

The primary controls on a microphone preamp are:

- **Gain** (may be called *Level* or *Trim*) controls how much the microphone signal is amplified. Most mic preamps have about 60dB of gain (which amplifies the mic signal a million times), but some have as much as 80dB to accommodate low-output ribbon microphones or field audio recording, where the signals captured by the mic are extremely quiet.

- **Metering** on a mic preamp can be something as simple as a single indicator that signals an overload, to a full-on ladder-style LED peak meter as found on consoles or DAWs. Many times there is a combination, with the overload indicator at the top of the full meter. (See below for setup tips using the metering.)

- **Input Pad** (or *attenuator*) is usually a switch that attenuates the signal from the microphone from 10 to 20dB (it's different for every mic pre) to keep the preamp circuitry from overloading.

- **Phase** changes the polarity of the microphone signal due to either a misplaced or a mis-wired microphone. Set the switch to the position that has the most low end. (For more on phase, see Chapter 5, "Microphone Placement Fundamentals.")

- **A high-pass filter** (sometimes called *low-cut*) attenuates the low frequencies at anywhere from 40Hz to 160Hz to eliminate unwanted low-frequency noise (such as truck rumble). On most preamps the frequency is fixed, but many have a variable frequency control.

- **Impedance:** A few mic preamps feature an Impedance control, which is used to properly match the impedance of a microphone to the preamp. This is less important today than back in the days when audio equipment required a precise 600-ohm load in order to operate within specifications (from the '40s through the '70s). Experiment with the various settings and select the one that has the fullest sound with the most low end.

- **Output:** Some preamps have an additional parameter that controls the overall output of the device; it's usually labeled *Output*. This comes in handy because it's an additional safeguard against distortion by allowing you to control the level going into the next device in the signal chain. Adjusting the *Output* control can help provide the best signal-to-noise ratio with the least amount of distortion. It also allows you to get different sounds from the preamp in conjunction with the Input control. By raising the input and lowering the output, you can intentionally introduce an amount of saturation or distortion into the signal.

Setup Method

Here's a recommended way to set up a microphone preamp, regardless of the manufacturer or model.

1. Adjust the gain until the clip light just flashes only on the loudest sections of the recording, then back it down until it no longer flashes even on the loudest peaks. If the preamp doesn't have an output control, adjust the gain control until the following device's meter is in the -6 to -10dB range. This gives you the best combination of low noise with the least distortion (unless of course you like distortion, in which case you want the overload indicator to remain on much of the time). If you set the gain of the mic amp too low, you might have to raise the gain at another place in the signal chain, which can raise the noise to an unacceptable level.

2. If the input signal is so hot that the overload indicator lights even if the gain control is turned all the way down, select the input pad if the unit has one. Now increase the gain control back to just before the overload lights, as described in Step 1. Even if the gain control is set low but not completely off, you still might want to insert the pad and readjust the gain control to be certain that it will not overload, and to give you more precise control over the gain.

TIP: If the preamp doesn't have a built-in input pad, you may have to resort to what's known as an inline pad, which is inserted at the end of the microphone cable (see Figure 3.19).

Figure 3.19: An inline input pad
Courtesy of Shure

3. If the preamp has an output control, set it so that the next processor in line is not overloading. Adjust the output control of the preamp so that the meter reading on the following device (probably the DAW) is in the -6 to -10dB range. You can also use the combination of the gain and output controls to get different sounds from the preamp. A setup as described above will provide a clean sound, while a setting with a higher input setting and a low output setting will provide a more aggressive sound.

4. For most recordings, insert the high-pass filter (if the preamp has one) to eliminate the low frequencies that add nothing to the sounds that are being captured. The exceptions are instruments such as the kick drum, floor tom, or bass, where capturing all the low-frequency information is desirable.

5. If you are recording with multiple mics, select both positions of the phase switch and select the one that provides the most low end. You may hear no change at all, in which case you should leave it in its deselected position. The same applies when there is only one mic used on the recording.

6. If the preamp has an impedance switch, try the different selections and choose the one with the most body and low end.

DIRECT INJECTION

Direct injection (DI or going direct) of a signal means that a microphone is bypassed, and the instrument (always electric or electrified) is plugged directly into the console or recording device. This was originally done to cut down on the number of mics (and therefore the leakage) used in a tracking session with a lot of instruments playing simultaneously. However, a DI is now used because it either makes the instrument sound better (as in the case of electric keyboards) or is just easier and faster. Plus it also performs the duties of taking an unbalanced 1/4 inch instrument cable and making it compatible with a balanced XLR mic input.

Why can't you just plug your guitar or keyboard directly into the mic preamp without the direct box? Most preamps now have a separate input dedicated for instruments, but there was a time when that wasn't the case and plugging an electric guitar (for instance) into an XLR mic input would cause an impedance mismatch. This would change the frequency response of the instrument (although it wouldn't harm the instrument or preamp), usually causing the high frequencies to drop off and therefore make the instrument sound dull.

Advantages Of Direct Injection

There are a number of reasons to use direct injection when recording:

- Direct-box transformers provide ground isolation and allow long cable runs from high-impedance sources such as guitars and keyboards without excessive bandwidth loss.
- The extremely high impedance of the DI ensures a perfect match with every instrument pickup to provide a warmer, more natural sound.
- The length of balanced output cable can be extended to up to 50 feet without signal degradation.
- Converts an unbalanced 1/4 inch instrument output into a balanced XLR mic input.

Direct-Box Types

There are two basic types of direct boxes: active (which contains electronics that adds gain and requires either battery or AC power), or passive (which provides no gain and doesn't require power). Which is better? Once again, there are good and poor examples of each. Generally speaking, the more you pay, the higher quality they are (see Figures 3.20).

Figure 3.20: Demeter Active VTDB-2b direct box
Courtesy of Demeter Amplification

An active DI sometimes has enough gain to be able to actually replace the mic amp and connect directly to your computer's audio interface.

You can build an excellent passive DI around the fine Jensen transformer specially designed for the task (see jensen-transformers.com for do-it-yourself instructions), but you can buy basically the same thing already built from Radial Engineering in their JDI direct box (see Figure 3.21). Also, most modern mic pres now have a separate DI input on a 1/4-inch guitar jack.

Figure 3.21: Radial JDI direct box
Courtesy of Radial Engineering

Direct Box Setup

Not much setup is required to use a direct box. For the most part, you just plug in the instrument and play. About the only thing that you might have to set is the gain on an active box (which is usually only a switch that provides a 10-dB boost or so) or the ground switch.

Most DIs have a ground switch to reduce hum in the event of a ground loop between the instrument and the DI. Set it to the quietest position.

Amplifier Modelers

The amplifier emulator, which is basically a glorified active direct box, has now been around for some time and has become a staple in so many studios (see Figure 3.22). A modeler attempts to electronically duplicate the sound of different guitar and bass amplifiers, speaker cabinets, and even miking schemes, including pedal and effects setups.

The advantages of these boxes are that they're quick and easy to set up, they give a very wide tonal variation, and they provide the proper interface to just about any analog or (in some cases) digital recording device. Models by Kemper, Fractal, Line 6, Neural, as well as a variety of pedals, now can sound remarkably like a properly miked vintage amplifier in a great studio through a terrific signal chain, which is why many guitarists now opt for one instead of the real thing (especially considering the price).

Figure 3.22: The Line 6 Helix LT Amplifier Modeler
Courtesy of Line 6

COMPRESSOR/LIMITERS

A compressor/limiter is frequently inserted into the microphone signal chain for two reasons: for dynamics control to prevent a signal overload, or to change the tonal characteristics of the sound.

Primary Controls

The primary controls on a compressor are:

- **Threshold** sets the signal level at where the automatic gain reduction begins. Below that level, the compressor does nothing. When the input level rises above the threshold point, the compressor reduces the volume automatically by the amount set by the *Compression Ratio* control.

- **Attack Time** determines how quickly the volume is reduced when the input exceeds the threshold. If set too slow, then the signal transients won't be acted on and an overload may occur. To catch the transients of the signal, a faster attack time is selected. If set too fast, the transients will be cut off and the sound will be dull and lifeless. Then again, sometimes the sound you're looking for lets those transients through on purpose.

- **Release Time** determines how quickly the volume returns to normal after being reduced. If set too fast, this change becomes audible as the volume quickly swings up and down (this is called *pumping* or *breathing*.) Setting the release time longer eliminates this.

- **Compression Ratio** determines the amount of compression that will occur. A setting of 1:1 (1 to 1) does nothing, except maybe color the sound a bit due to the electronics of the compressor. A setting of 2:1 means that if the input rises 2dB above threshold, the output level will increase by only 1dB. A setting of 10:1 means that the input must climb 10dB above the threshold before the output increases 1dB.

- **Sidechain:** Many compressors also have an additional input called a *sidechain input*, which is used for connecting other channels to it (see Figure 3.23). The level from that channel then triggers the compressor to lower the level. A good example would be a signal from a kick drum triggers the compressor sidechain of a compressor on the bass so when the kick hits, the bass lowers in level. A side chain isn't needed for normal compressor operations, so many manufacturers choose not to include it on their units.

Figure 3.23: A compressor side-chain input and output
© 2023 Bobby Owsinski

- **Bypass:** Most compressors, especially most of the plug-in versions, have a Bypass control that allows you to hear the signal without any gain reduction taking place. This is useful to help you hear how much the compressor is controlling or changing the sound, or to make it easy to set the Output control so the compressed signal is the same level as the uncompressed signal.

Types of Compressors

There are four different electronic building blocks that may be used to build an analog hardware compressor. These are:

- **Optical:** A small light bulb and a photocell are used as the main components of the compression circuit. The time lag between the bulb and the photocell gives it a distinctive attack and release time (such as in a Teletronix LA-2A).

- **FET:** A solid-state field-effect transistor (FET) is used to vary the gain, which has a much faster response than an optical circuit (a Universal Audio 1176 is a good example).

- **VCA:** A voltage-controlled amplifier (VCA) circuit is a product of the '80s and has both excellent response time and much more control over the various compression parameters (the dbx 160 series is an example of a VCA-type compressor, although some models didn't have a lot of parameter controls.)
- **Vari-Gain:** The vari-gain compressors are sort of a catch-all category because they use other ways to achieve compression besides the first three. These are usually tube-based (such as the Fairchild 670 and Manley Vari-Mu).

As you would expect, each of the above has a different sound and different compression characteristics, which is the reason why the settings that work well on one compressor type won't necessarily translate to another. Modern plugin compressors can easily emulate any one of the above types.

Limiting

While a compressor increases the low signal levels and decreases the loud ones to even out the dynamic range, a limiter keeps the level from ever going much louder once it hits the threshold. Compression and limiting are closely related, the main difference being the setting of the ratio parameter and the application. Any time the compression ratio is set to 10:1 or greater, the result is considered limiting, although most true limiters have a very fast attack time as well.

A limiter is essentially a brick wall for level, allowing the signal to get only to a certain point and little more beyond. Think of it as doing the same thing as a governor that's sometimes used on 18-wheel trucks owned by a trucking company to make sure that they're not driven beyond the speed limit.

Once you hit 55 mph (or whatever the speed limit in your state is), no matter how much more you press the gas pedal, the truck won't go any faster. The same theoretically occurs with a limiter. Once you hit the predetermined level, no matter how much you try to go beyond it, the level pretty much stays the same.

Most modern digital limiters (either hardware or software) have a function known as *look ahead*, which allows the limiter to look at the signal a millisecond or two before it hits the detector circuit in the limiter. This means that the limiter acts extremely fast and just about eliminates any overshoot of the predetermined level, which can be a problem with analog limiters because they react much more slowly to transients.

Limiting is used a lot in sound reinforcement for speaker protection (there are limiters on some powered studio monitors as well), but also during recording. Many engineers will place a limiter on an instrument like drums, bass, or vocals that have a lot of transients, just to make sure that a sudden shout or hit doesn't cause an overload. In this case, the amount of limiting used is very light (maybe only a dB or two at most).

Typical Compressor/Limiters

As with microphones and preamps, the vintage units of the past are still highly desirable. Here are a few famous models that are frequently used during tracking.

Teletronix LA-2A

The Teletronix LA-2A is perhaps the most popular of all tube compressors. Early LA-2As can be identified by their gray faceplate, while later models feature a brushed aluminum face with a switch on the back to swap between limiting and compression functions (in the earliest models, this was accomplished through internal jumpers). While not at all sonically transparent, the LA-2A provides an airy sound (especially on vocals) heard on literally thousands of hits. There are only three controls: an input threshold control, a makeup gain control, and a Comp/Limiter switch that changes the compression ratio, but they're quite enough to do the job very well (see Figure 3.24).

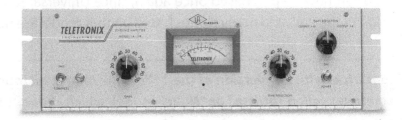

Figure 3.24: An LA-2A
Courtesy of Universal Audio

There are numerous clones of the LA-2A available on the market, as well as a reissue of the unit by Universal Audio, the descendant of the original manufacturer.

dbx 160A

dbx made a series of VCA-style compressors that are still found in studios everywhere. These include the original 160 to the later 160x to the most recent 160A (see Figure 3.25). Unfortunately, the 160A has been discontinued, but you can still get that unique compression in a 500 series module called the 560.

Figure 3.25: A dbx 160a
© 2023 Bobby Owsinski

UREI 1176

Introduced in 1967, the 1176 has become one of the most storied pieces in audio history, being a staple in nearly every studio since. Released in eight versions (from A to H), the earliest 1176A and B models were silver-faced with a blue stripe around the meter and featured push-button ratio selection of 4:1, 8:1, 10:1, 20:1, as well as attack and release controls. These attack and release controls actually work backwards, with the higher numbers (7 is highest clockwise) being faster than the lower numbers (see Figure 3.26).

Figure 3.26 UREI 1176
Courtesy of Universal Audio

Of all the versions, the model D or E variants are the most desirable. The 1178 is a stereo/dual-mono version of the 1176 with single controls for both channels.

New versions of the 1176 can now be purchased once again, since Universal Audio has reissued the black-face model E version of the unit, and many clones based on that model are now available as well.

TIP: One of the neat tricks with any version of the 1176 is pushing in all four ratio buttons simultaneously (sometimes called "British" or "Nuke" mode), which changes the compression circuitry so it increases the harmonic distortion and has a very aggressive sound. It makes the meter go crazy, but it sure sounds cool.

Empirical Labs EL 8 Distressor

Released in 1996, the Distressor immediately gained traction in studios everywhere thanks to what it calls digitally-controlled analog knee compression. It has the unique ability to sound similar to the other compressors mentioned above but with even more color thanks to its unique Distortion and Harmonics modes (see Figure 3.27).

Figure 3.27 Empirical Labs EL8 XDistressor
Courtesy of Empirical Labs

Compressor/Limiter Setup

For dynamics control during recording, the compressor is usually set to a higher compression ratio (an 8:1 or higher gain ratio) and set so that the signal doesn't exceed a certain level (usually just before clipping or distortion). Usually only a few dB of compression is needed. In other cases, such as with voice-overs, the unit may be used to smooth out the differences between words and phrases and is set to a low compression ratio (2:1 or higher but less than 10:1) with as much as 6 to 8dB used.

There are two things that are important in setting up a compressor: the timing of the attack and release, and the amount of compression. Here are a few simple steps to help you set one up. *Remember: The idea is to make the compressor breathe in time with the song.*

1. Set the *Threshold* control until there is a few dB of compression. The amount isn't as important as your ability to hear it during the initial setup.

2. Set the *Ratio* control to a high compression ratio (8:1 or higher) if you want to control peaks, and a lower compression ratio (2:1 to 6:1) if you want to smooth out the dynamics. *Keep in mind that the higher the ratio is set, the more likely you will hear the compressor working.*

3. Set the attack time as slow as possible and the release time as fast as possible. The amount of compression will drop to zero when the controls are set this way, so increase the *Threshold* control until you see compression occurring again.

4. Turn the *Attack* faster until the instrument begins to sound a bit dull. (This happens because you're compressing the transients in its sonic envelope.) Stop increasing the attack time at this point and even back it off a little. Set the attack time a little faster if you want to control the peaks, or a little slower if you want the sound of the transients to be preserved. If you don't like the sound when the peaks are compressed, try another compressor.

5. Adjust the release time so that the volume goes back to 90 to 100 percent normal with the pulse of the song. In other words, the release should timed so the instrument breathes with the pulse of the song.

TIP: When in doubt, set the attack and release times to midway and leave them there.

---

6. With the attack and release set, adjust the *Threshold* control for the desired amount of compression or limiting.

7. Select *Bypass* to see whether the level has changed. Compression or limiting should automatically decrease the output of the signal, so adjust the *Output* control until the level is the same regardless of whether the *Bypass* is selected. Adjust the *Output* control further if more gain is required or lower if the next component in the signal path is overloading.

Basic Recording Equipment

DAW RECORDING

Most musicians have some sort of recording capability at home these days, since the technology is so inexpensive and readily available. In fact, it's now easy to put together a studio with incredible power and capabilities that go far beyond what was possible only a short time ago. That said, most of us now record directly into a computer, so there are a few important concerns that you need to address before you hit the Record button.

COMPUTER RECORDING

Back when recording consoles and tape machines where being used (as well as in those studios that still have them), recording was sometimes actually easier than what you find with a typical DAW these days. Once you patched everything together, you could get right down to working. Today it's more than likely that you'll end up trying to track down why your DAW isn't recording or why you're not hearing anything even though it's clear that you're recording. The software that we use brings a whole new set of tech challenges with it.

While we can't address each particular digital audio workstation package in this book, we can address some commonalities between them all.

The Computer Audio Interface

When it comes right down to it, the make and model of computer interface that you use has just as much bearing on the sound you capture as the mics and preamps. This comes down to two devices embedded in most interfaces: the analog-to-digital convertor (often shortened to ADC), which is the unit that converts the analog signals into the digital language of the DAW, and the digital-to-analog convertor (often shortened to DAC), which is the unit that converts the digital signal back to analog for playback.

Back when computer-based audio workstations were new, these units were separate pieces, but today most DAW interfaces by manufacturers such as Digidesign, Focusrite, Universal Audio, and MOTU have ADCs and DACs that are built into the same box along with a few mic preamps. That said, it's still true that when you get into higher-end systems such as Pro Tools|HD, the ADCs and DACs come in separate higher-quality units capable of greater fidelity. You'll even find specialized super high-end outboard units like those from Apogee, Burl or Lavry capable of even higher fidelity than those used in DAW manufacturer's interfaces.

The good news is that most convertors used today, even in budget all-in-the-same-box interfaces, sound really good compared to the high-end units that were available in the early 2000s when digital recording was in its infancy. That, coupled with the fact that most DAW software such as Nuendo, Cubase, Logic, Cakewalk, or Ableton for instance, now allows you to mix and match the hardware to the level of quality that you need and can afford without much hassle, which means that for the most part we're no longer limited to recording inferior audio due to the interface.

That said, a higher price usually does buy you increased audio quality, but the sound quality still depends upon the weakest link in your signal path.

Monitor Latency

Latency is a measure of the amount of time (in milliseconds) for the audio signal to pass through the DAW during the recording process. This delay is caused by the time it takes for your computer to receive, understand, process, and then send the signal back to the audio outputs. We used to worry about this a lot more in the past than we do today, thanks to the increased computer horsepower available, but it's still something that rears its ugly head from time to time. The good news is that once you find the correct computer settings, you don't have to mess with them too often afterward.

Here's the problem: If you have what's known as high latency, you hear a note way after you play or sing it, which is something you definitely want to avoid. High latency means it takes too long for the audio input to become audio output, so there's a lag time between the time you play or sing a note and when you hear it. A very small lag time of 3 to 6 milliseconds is tolerable (although not ideal), but anything beyond that creates everything from a phasing sound to a full echo, which makes it anywhere from distracting to impossible to sing or play with.

Different Ways To Lower Latency

There are four systems for lowering monitor latency - native (in software), analog, DSP, or hybrid. The differences between them is how low the latency can be, if plugins can be used, and cost. Let's look at them.

Native Monitoring

A native system uses the DAW's software to control the amount of latency that you'll hear.

As you lower your latency, the music that you're recording stays more in sync with the music that you're playing back, up to a point. If you try to set the latency parameters too low, the audio stream can break up into static, noise, or distortion, since the computer doesn't have the time to process it all.

Still, it is possible to adjust the latency in a native system as low as it can go without causing the computer to stutter, which is accomplished either through your DAW or interface settings, or through third-party audio drivers.

The parameter that most computer audio interfaces use to set the latency is called the *Input Buffer*. The smaller the buffer setting, the lower the latency becomes, but the harder the CPU has to work. If you lower the buffer size too much, the setting can produce crackling noises, which can happen even with a computer with a lot of horsepower. These noises crop up when the CPU literally has to drop audio bytes because it can't keep up with the audio stream.

Today's fast computers can get the I/O buffer size down to around 64 samples (which means a 1.3-ms latency at a 48-kHz sampling rate) without too much trouble, but the more tracks and processing you add (especially when running at sampling rates higher than 48 kHz), the harder the computer's CPU will have to work, which means you may need to increase the buffer size, and therefore the latency, to prevent dropouts.

Other things that can affect latency but are usually not thought of are parameters such as your WiFi connection, how many programs are running, whether your computer has a virus, current computer background activity, connected yet unneeded peripherals, and even the graphics card. When in doubt, *turn it off!*

In the end, how low your latency can be set is dependent upon such factors as computer speed, system bus speed, audio interface performance, and system memory. Most computers purchased today are powerful enough that you can get the latency low enough to be usable, but you still have to experiment to find the settings that provide the best performance.

TIP: It's best to not use any software audio processing, such as compressors or EQ, when recording using native monitoring because each plug-in adds anywhere from a little to a lot of latency. Keep the signal path as efficient as possible with as few things between the mic and the recorder as you can, and your signal will not only sound better but will stay in sync as well.

Analog Monitoring

Many audio interfaces today come equipped with what's known as zero-latency monitoring, which is an analog buss that loops directly from the interface's input to its output without passing through the computer. Once you've set up this routing either on your interface's front panel or control panel applet (it comes with the interface), the player or singer will be able to monitor the backing tracks and get his or her performance in sync without any time delay. The downside is you're not able to use any audio plugins.

If your interface doesn't have zero-latency monitoring, you can accomplish the same thing if you have an analog or digital hardware console. Just connect the interface input to an aux, bus, direct, or digital output of the mixer to avoid recording the entire temp mix into the new track. While there is some latency in digital mixers, it's kept very low (2 ms or less) thanks to an operating system that's specifically optimized for the task.

DSP Monitoring

In order to lower the latency without worrying about computer settings, manufacturers like Avid and Universal Audio have DSP (Digital Signal Processing) solutions available. That means that there are dedicated DSP chips either on hardware cards or built into the audio interface hardware.

This can get the latency down into the 1ms range, which means it's essentially gone, plus the DSP chips provide the ability to use a range of plugins while recording without adding to the latency.

While this sounds great, the big problem is the cost is higher than many home studio owners want to pay. Plus, you can only use the plugins available from the manufacturer.

Hybrid Monitoring

The hybrid system is the most recent advancement in the elimination of recording latency, where the processing is split between DSP and native mix engines. That means that your session runs as normal on your host computer, but when you need to record with near-zero latency the computer is supplemented with the extra DSP.

The advantage is that a high powered computer system is not required, even though the cost for the hybrid hardware may still be more than a simple audio interface. The workflow for using the system is also relatively transparent as compared with the more complicated DSP systems.

Right now Apogee Dual Path, Avid Carbon, Universal Audio Luna, and PreSonus StudioLive all feature hybrid monitoring systems.

> **TO LOWER LATENCY:**
> - Lower the input buffer size.
> - Use zero-latency monitoring, or...
> - Use an outboard console.
> - Avoid the use of plugins unless your interface has onboard processing.
> - Turn off programs and background activity on the computer.

USB versus Thunderbolt

Newer computer audio interfaces, all recent MacBooks, Chromebooks and Windows laptops, Android phones and Apple iPads (and soon iPhones) use the same connector that, depending on the circumstances, can be either USB-C or Thunderbolt. Since both of these interface technologies use exactly the same connector, it's easy to get confused about the difference between them and what it means in real world use. It should be noted that the connector has no right side up and can be inserted either way (see Figure 4.1).

Figure 4.1: Thunderbolt 3 and 4, and USB-C cable connector
© 2023 Bobby Owsinski (Source: Pixabay)

To make matters even more confusing, there are different flavors of both Thunderbolt and USB that muddy the water even more. Here's a chart that explains some of the differences.

Table 4.1: Interface Speeds

INTERFACE	SPEED	POWER	DISPLAYS	CONNECTOR
Thunderbolt 4	Up to 40Gbit/sec	Delivers up to 100 watts	Up to two 4k monitors or one 8k	Type C
Thunderbolt 3	Up to 40Gbit/sec	Delivers up to 100 watts	One 4k monitor	Type C
USB-C	Up to 20Gbit/sec	Delivers up to 100 watts	One 4k monitor	Type C
USB4	Up to 40Gbit/sec	Delivers up to 240 watts	One 4k monitor	Type C

As you can see, Thunderbolt and USB4 are the fastest connections available, which means that in theory they can easily stream far more audio channels than you'll ever need, but they also excel in high-bandwidth situations such as video editing as well.

That said, USB-C, USB4, and either flavor of Thunderbolt are somewhat interchangeable, with the only difference being the charging power, data transfer rate and the number of high-resolution monitors that it can connect to.

Cables

The cables that you use to connect a modern computer peripheral to the computer can make a big difference in the performance. First of all, the shortest cables provide the highest transfer speeds. This means that the length is generally limited to around 1 meter (3.3 feet) long in order to hit the highest specs.

USB-C and Thunderbolt cables are different as well, with the connector ends designating which technology each cable is intended for (Thunderbolt cables carry a lightning bolt logo - see Figure 4.2).

While both are essentially interchangeable and will work in many situations, if you want the highest performance then stay with the assigned cable.

Figure 4.2: A Thunderbolt 3 connector
© 2023 Bobby Owsinski

Cables can be active (with a microchip built into the connector that needs power) or passive. Active cables have the advantage of being longer while still maintaining the highest performance, but are also more expensive.

The best advice is to use the cable that came with your audio interface or hard drive to get the performance you're expecting. If a longer one is required, contact the manufacturer for a recommendation.

Legacy Formats

Many older computer audio interfaces use either the now obsolete Thunderbolt 1 or 2, Firewire, or even a USB 3.0 connection. Thunderbolt 1 and 2 each have different connectors and are limited to a top data speed of 10Gb/sec. Firewire also comes in two flavors; Firewire 400 and 800, while USB has so many variations that it's confusing even to people who work with it every day.

The idea with all these different connection technologies is to increase the speed of the digital data transfer so you can have more tracks available in your DAW, or transfer computer data faster to and from storage. Except for real power users, Firewire and Thunderbolt 1 and 2 connections are plenty fast for most audio applications, but the hassle of using them with a recently made computer can be surprisingly time consuming and expensive because of the need for various cable adapters.

The bottom line is that adapters and docks are available to make sure that you can still connect to those earlier interfaces if you must, but that technology is fading away fast. The few interfaces that used either USB 3.0 are certainly serviceable but are usually limited to recording just a few tracks at a time.

Sample Rate

Sample rate is one of the factors that determines the quality of a digital audio signal. The amplitude of the analog audio waveform is measured by the analog-to-digital converter at discrete points in time, and this is called sampling. The more samples per second of the analog waveform that are taken, the better the digital representation of the waveform that occurs, which results in wider frequency bandwidth (see Figure 4.3). This is known as the *Sample Rate*.

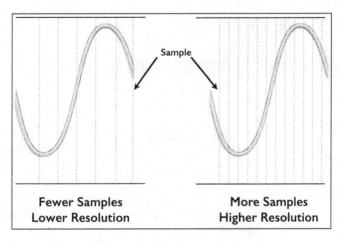

Figure 4.3: Sample rate
© 2023 Bobby Owsinski (Source: iStock Photo)

Audio on a CD has a sample rate of 44,100 times a second (or 44.1kHz), which, thanks to a law of digital audio call the Nyquist Theory, yields a maximum audio bandwidth of about 22kHz, or half the sample rate. A sample rate of 96kHz gives a better digital representation of the waveform because it uses more samples, and yields a usable audio bandwidth of about 48kHz. A 192kHz sample rate provides a bandwidth of 96kHz.

TIP: *The higher the sampling rate, the better the representation of the analog signal and the greater the audio bandwidth will be, which means it can sound better!*

Although a higher sample rate yields a better representation of the analog signal, some listeners can't always tell the difference due to the the limitations of their playback system, their listening environment, their signal path, or even the type of music they listen to. Even though the CD sample rate is 44.1kHz, 96kHz has become the standard for music recording, thanks to Apple's iTunes and now Apple Music encouraging the submission of high-resolution files.

The downside of a higher sample rate is that takes up more digital storage space, with 96kHz taking up twice as much as 48k and 192k taking up twice as much again as 96k. That's no longer much of a problem, though, as storage devices today are massive compared to the needs of a typical song.

Table 4.2: Typical Sample Rates

TYPICAL SAMPLE RATES	COMMENTS	CAVEATS
44.1kHz	The CD sample rate	Fewer CDs being made, so the minor advantage of using this sample rate is lost. Lowest professional digital resolution.
48kHz	Standard for film and TV	Lowest recommended sample rate.
96kHz	High-resolution standard	Most pro recordings today done at 96kHz, recommended master delivery for many streaming services. Takes up twice the storage space of 48kHz.
192kHz	Audiophile	Only half the channels and plugins available at 96kHz on some DAWs. Many plugins don't operate. Takes up twice the storage space of 96kHz.

Bit Depth

The digital word length is the other factor in audio quality, and it's somewhat the same as sample rate in that more is better. The more bits in a digital word, the better the dynamic range, which means it sounds better. Every bit used equals 6dB more dynamic range. Therefore, 16 bits yields a maximum dynamic range of 96dB, 20 bits equals 120dB, and 24 bits provides a theoretical maximum of 144dB. From this you can see that a high-resolution 96kHz/24-bit (usually just abbreviated 96/24) format is far closer to sonic realism than the current CD standard of 44.1kHz/16-bit.

Today, most recording is done at 24 bits, as there's no advantage to using less. While back in the early days of digital recording hard-drive space was at a premium, that no longer applies. That said, both CD and many online formats are limited to 16 bits, many streaming services with high-resolution tiers like Apple Music and Tidal now encourages 24-bit delivery regardless of the sample rate.

TIP: The longer the word length (the more bits), the greater the dynamic range and therefore the more realistic it sounds.

---

32 Bit Recording

In the recording studio a competent engineer can be reasonably sure that once the levels are set, even the loudest part of the song won't exceed 0dB Full Scale (the highest the signal can go) and therefore distort the signal. That goal is a lot more difficult to achieve in field recording however, since one never knows when an actor might scream or an unexpected stage prop will make an incredibly loud noise. When that happens, 0dB FS can be quickly exceeded causing distortion, and it can't be fixed later. Many field recorders are now incorporating what's called a 32 bit float format, which alleviates this problem.

A 32 bit float system eliminates the need for setting an input level as overload distortion is just not possible. While 24 bit systems have a dynamic range of 144dB, this system has an incredible 1,528dB, plenty to accommodate even the loudest sound on the set. The recorded levels may appear to be either very low or very high while recording, but they can easily be scaled after recording by the DAW software with no additional noise or distortion. If a recording appears to be mostly distorted with the overload indicator lit most of the time during playback, all that's needed is to lower the level until it looks normal, and the signal will sound fine with no distortion.

The downside is that a 32 bit float recording takes up about a third more storage space than a 24 bit recording, but once again, that's not a problem these days as computer storage is cheap. Some recorders that employ this technology include the Zoom F3 and F6, and the Sound Devices MixPre-3 II.

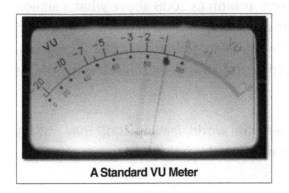

Figure 4.4: An old-fashioned VU meter
© 2023 Bobby Owsinski

SETTING THE RECORDING LEVEL

In the old analog days, it was relatively easy to set the recording level, since everyone knew that the level on the VU meter (see Figure 4.4) should be kept out of the red area above 0dB. Now we mostly live in a digital world, and except in rare cases, VU meters are a thing of the past. That means that we're dealing with mostly digital peak level meters, which require a whole different mindset when it comes to the proper recording level. But just how high should that level be?

Believe it or not, you *don't* have to record at a level close to 0dB FS (the highest it will go before the red overload indicator lights) these days. In the early days of digital recording, this practice was a necessity in order to keep the noise to a minimum, but modern 24-bit recording no longer has this limitation. That means if your signal peaks at somewhere between -6 and -10dB or even lower on the channel meter, it will usually sound fine, but that brings up the need to be aware of the concepts of headroom and gain-staging when it comes to setting your signal levels for the cleanest audio.

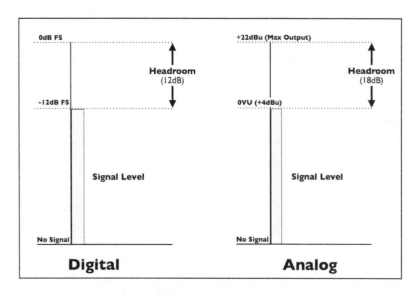

Figure 4.5: Measuring headroom
© 2023 Bobby Owsinski

Headroom

Headroom means the amount of level left above the average signal level before it distorts. For instance, if you're recording at -12dB and you're looking at a typical digital peak meter, you have 12dB before the signal will overload, or 12dB of headroom (see Figure 4.5). Many of today's analog mixers overload at +22dBu, which means that if the signal is hitting 0dB on a typical analog VU meter, the output of the console is at +4 dBu (see the glossary for why), which means there's 18dB of headroom (22 minus 4 = 18).

The reason why we want and need headroom in the first place revolves around trying to capture the cleanest audio possible. The more headroom you have, the more likely that you'll be able to record signal transients from instruments such as tambourines, drums, and percussion that are so fast that a VU meter can't track them. These transients can typically range as high as 20dB above what a analog VU or digital RMS meter might be telling you (peak meters are much closer to the actual true recording level), and recording too hot means those transients cause the signal to go into overload, if only for a millisecond or less. This results in not only a slightly dull recording, but one that sounds less realistic as well. The solution is more headroom.

While leaving 18dB for headroom might be excessive in the digital world, leaving 10dB may not be. Since it's easy enough to make up the gain later, you won't increase the noise, and your mix will be cleaner. Why not try it?

TIP: Recording between -12 and -6dB FS provides enough headroom with an excellent signal-to-noise ratio.

Gain Staging

Gain staging means setting the proper level at each section of the signal path so that no one section overloads and the noise remains at its lowest. On either an analog or digital console (even if it's virtual in a DAW), proper gain-staging makes sure that the microphone doesn't overload the input stage, the output of the input stage doesn't overload the equalizer section, which then doesn't overload the panning amplifier, which then doesn't overload the fader buffer, which then doesn't overload the buss, which then doesn't overload the master buss. This is the reason why a pre-fader and after-fader listen

(PFL and AFL) exists: to monitor as many sections in the signal path as possible to make sure there's no distortion occurring, and if there is, to enable you to find it easily.

While all of these stages are tweakable, one rule exists in the analog world that aptly applies in the digital world as well.

> **The level of the channel faders should always stay below the subgroup or master fader.**

This means that the level of the master fader should always be placed higher than each of the channel faders (see Figures 4.6, 4.7, and 4.8) to be sure that an overload doesn't occur. While one or two channels might be okay if placed slightly above the master or subgroup fader (it's almost inevitable in every mix), just a single channel with big chunks of EQ (such as +10dB) or an insert with an effects plugin that's maxed can destroy any semblance of a good-sounding mix.

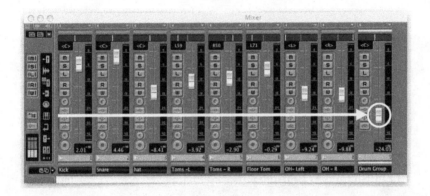

Figure 4.6: Channel faders too high, subgroup fader too low
© 2023 Bobby Owsinski (Source: Steinberg Media Technologies Gmbh)

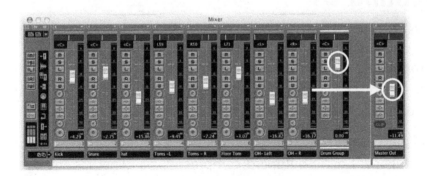

Figure 4.7: Subgroup too high, master fader too low
© 2023 Bobby Owsinski (Source: Steinberg Media Technologies Gmbh)

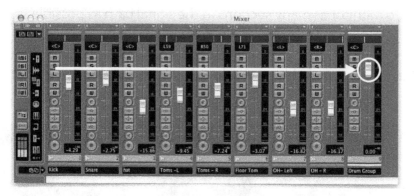

Figure 4.8: Channel and subgroup faders at correct levels
© 2023 Bobby Owsinski (Source: Steinberg Media Technologies Gmbh)

Because many consoles have plenty of headroom these days, this principle hasn't been religiously followed, but it has been a golden rule since day one of modern recording consoles.

Not following this rule is the main reason for lack of fidelity and power in the DAW, however. *The master buss is overloaded!*

> **RULES FOR GAIN STAGING**
> 1. Keep all channel faders below the subgroup or master fader. Keep all subgroup faders below the master.
> 2. When using large amounts of EQ or a plugin with gain, lower the channel fader rather than bringing up the others around it.
> 3. Leave lots of headroom by recording at a lower level. It's easy to raise the level later when you're mixing.

Troubleshooting

If something sounds distorted, use the following steps to track it down:

1. **Is the microphone preamp overloading?** Is the red overload indicator lit or is the meter peaking into the red? If so, decrease the input gain or select the input pad or the pad on the mic, if it has one.

2. **Is the signal path overloading somewhere else in the signal chain?** If you're using a console (regardless of its size) or an outboard compressor, see whether any overload indicators are lighting or if the meters are peaking into the red. If they are, decrease the output level of the stage or processor before the overload.

3. **Is your DAW overloading?** Be sure that no overload indicators are lit. This shouldn't happen if you keep your input level between -6 and -10dB or lower. If it does, decrease the input level on the DAW or the output level of the previous gain stage or processor.

4. **Is your playback signal path distorting?** It may not be what you're recording that's causing the problem, it could be in the playback signal path causing the distortion instead. If you're listening through a console, check to see whether one of the channels is overloading. Are the monitors turned up too loud? Many monitors have a built-in limiter to keep them from blowing, but this comes at the expense of the audio quality. Are there any overload indicators lit anywhere in the signal path? If there are, decrease the level from the DAW first, or turn down any input level controls in the playback signal path.

5. **Could the problem be a mic or cable?** Try replacing the cable first, since that's the most likely place to have a problem. If the sound isn't any cleaner, try a different mic.

6. **Check to see whether something in the room is rattling.** Sometimes the mics pick up rattling from speaker cabinets or extra drums on the floor that sound just like something is distorting. For instance, the buzz that comes from a loose Marshall cabinet handle is notorious for sounding like distortion. Walk around the recording area and listen to the instrument and the environment closely, but be sure to have the player play the exact same part as when you heard the distortion. Sometimes the sound will only come from a single note, so by playing the same part you ensure that it can happen again so you can track it down.

MICROPHONE PLACEMENT FUNDAMENTALS

Recording can be broken down into a number of fundamental steps. Follow these fundamentals, and you're more likely to capture the natural sound of the instrument or vocal, or any sound source for that matter, that you're trying to record. Of course, sometimes you're not looking for a natural sound and want to experiment, but being aware of the fundamentals will allow you to get to that point faster.

We can break down the very basics of mic placement into a few simple exercises: choosing the recording environment, choosing the mic, placing the mic, and then avoiding the destroyer of good sound - phase cancellation. Let's take a look.

THE FORMULA FOR GETTING GOOD SOUNDS

Contrary to what many who are starting out in recording might think, just having great recording gear doesn't automatically guarantee a great sound. Because each situation, even within the same project, is unique, you can't really quantify how much each variable contributes to the ultimate sound that you capture. However, you can generally break it down to something like this:

- **The player and the instrument contribute about 50 percent to the overall sound.** Sometimes a little more, sometimes a little less, but always the greatest portion. Keep in mind that a great player can make an inferior instrument sing while the opposite isn't necessarily true.

- **The room contributes about 20 percent to the overall sound.** Even on close-miked instruments, the room is far more responsible for the ultimate sound than many engineers realize.

- **The mic position contributes about 20 percent to the overall sound.** Placement is really your acoustic EQ and is responsible for how the instrument or vocal blends with the rest of the track. A move of as little as an inch can sometimes make a bigger difference than you might think.

- **The mic choice contributes about 10 percent to the overall sound.** This is the final bit that takes a good sound and makes it great.

THE FUNDAMENTAL CHOICES

Before we even get to mic placement, there are two fundamental choices facing the engineer: choosing the place in the room where the instrument sounds the best, and choosing the right mic for the job. Once these are in place, we can move on to placing the mic. Let's take a look.

Choosing The Best Place In The Room

Where you place the instrument or vocalist in the room can make a big difference in how that instrument ultimately sounds. That said, when you're recording a group of players at the same time (especially a group that has a loud rhythm section), finding the best-sounding place in the room for each instrument is sometimes secondary to any leakage concerns and player sight lines. With a single player or small ensemble, however, finding the most complementary place in the room for the instrument is crucial. Try these following steps before you settle on your final room placement:

1. **Check the sound of the room first.** Do this by walking around the perimeter and center of the room while clapping your hands to find a place in the room that has a nice, even reverb decay. If you can hear a "boing" (a funny-sounding repeated ringing overtone that acousticians call "flutter echo") when you clap, then you'll also hear some of that sound when you record the instrument or vocal, so it's best to try to find another place in the room where the reflections will sound smoother. If you can't find a place without a boing, place the instrument where it causes the room to boing the least, and try putting some soft material, such as acoustic panels, carpet padding, or acoustic foam, on one of the side walls to cut down on the reflections. (Check out *The Studio Builder's Handbook* [Alfred Publishing] for more information on room treatment.)

2. **Try the area of the room where the ceiling height is the highest.** Low ceilings are usually the cause of some bad reflections, so the higher the ceiling is the better the room will sound, and the better chance your recording will sound pleasing as well. If the ceiling is vaulted, try placing the instrument in the middle of the vault and then move it as needed.

3. **Stay out of the corners.** The corner normally causes a phenomena known as bass loading, meaning that the low frequencies are reinforced, causing some low notes to boom. This can lead to sympathetic tom ringing and snare buzzing on the drum kit when you're tracking, as well as a lot of extra low end (although sometimes that can be beneficial).

4. **Don't get too close to a wall.** The reflections (or absorption if the wall is soft) can change the sound of the instrument, especially if it's very loud and omnidirectional, like drums or percussion. The middle of the room usually works best.

5. **Stay away from any glass.** Glass will give you a lot of reflections that can change the sound of the instrument. If placement near glass is the only area available, try setting the instrument and mics up at a 45-degree angle to the glass to send the reflections off at a different angle.

6. **Try putting a rug under the vocal or instrument.** A hard floor can sometimes be detrimental to the sound due to the reflections off of it, and a rug is good at stopping the worst of them from happening. On the other hand, sometimes the reflections from a hard floor can enhance the sound. Try it both ways and choose which works best.

7. **Try placing an amp or speaker cabinet on a chair or road case.** When the amp is lifted off the floor, there are fewer phase cancellations (see later in the chapter) so the sound will be more direct and distinct. Acoustic foam like Auralex placed underneath an amp or cabinet also works.

Choosing The Right Mic

While it's safe to say that most engineers rely on experience when choosing which microphone to use in a given situation, these are some things to consciously consider when selecting a microphone.

- **There's no one mic that works well on everything.** Just because you have what could be considered a "great" mic doesn't necessarily mean that it will be the best choice in all situations. There are times when the characteristics of that mic just don't match up with the instrument or vocal you're recording, and another mic will work better. In fact, sometimes even an inexpensive mic can work better than an expensive one.

- **Select a microphone that complements the instrument you'll be recording.** For instance, if you have an instrument that has a very edgy high end (it sounds very bright), you normally wouldn't want to choose a mic that also has that quality, since those frequencies will be emphasized. Instead, you might want to choose a mic that's a bit mellower, such as a ribbon. This is one of the reasons that a ribbon mic is often preferred on brass, for instance. In the case of an instrument with a lot of transients (like a triangle or tambourine), dynamic mics often work better than condensers.

- **Is the mic designed to be used in the free field or in the diffuse field?** Free-field means the sound that comes directly from the source dominates what the mic hears. Diffuse-field means that the room reflections play a large role in what the mic hears. Mics designed for free-field use tend to have a flat frequency response in the high frequencies, and as a result can sound dull when placed farther away in room from the sound source. Diffuse-field mics have a boost in the upper frequencies that make them sound flat when placed farther away. A good example of a diffuse-field mic is the esteemed Neumann M 50, which was meant to be placed somewhat away from an orchestra, so it has a high-frequency boost to compensate for the distance.

- **Select a mic that won't be overloaded by the source.** Some mics are sensitive enough that you must be aware of where they're used. You wouldn't want to put certain ribbon or condenser mics on a snare drum with a heavy-hitting drummer, for instance. Even some dynamic mics have little tolerance for high sound-pressure levels, so always take that into account.

- **Choose the right polar pattern for the job.** If leakage is a consideration, then choose a mic with the proper directional capabilities for the job. If a mic is flat on-axis (at the front), it probably will roll off some of the highs when it's 90 degrees off-axis (on the side). If it's flat 90 degrees off-axis, it may have a rising high end when it's on-axis.

- **Is proximity effect an issue?** If you intended to place the mic within 6 inches or closer from the source, will the bass buildup from the proximity effect be too much? If you think that may be the case, consider an omni pattern instead.

- **A large-diaphragm condenser mic is not necessarily better than small-diaphragm condenser.** Believe it or not, small diaphragm condenser microphones can sometimes capture the lower frequencies better, are generally less colored off-axis than large-diaphragm mics, and have a smoother frequency response. Large-diaphragm mics are a little less noisy, though.

FINDING THE OPTIMUM PLACEMENT

Quickly finding a mic's optimum position is perhaps the single most useful talent an engineer can have. Sometimes the search resembles questing for the Holy Grail as much trial and error can be involved. That said, you should always trust your ears first and foremost by listening to the musician in the tracking room, finding the sweet spot, and placing your microphone there to begin. If you don't like the resultant sound, then move the mic or swap it with another. EQ is the last thing you should touch.

TIP: Mics cannot effectively be placed by sight, which is a mistake that is all too easy to make (especially after reading a book like this). The best mic position cannot be predicted, it must be found.

How To Find The "Sweet Spot"

How you listen to an instrument in the studio is just as important as the act of trying to capture its sound. As good as many microphones are, they're still no match for our ears, and we can sometimes be fooled by what we hear over the monitor speakers. Here are a few tips to help you listen more closely to the way the mic you're using is capturing the sound.

- To correctly place an omni microphone, cover one ear and listen with the other. Move around the player or sound source until you find a spot that sounds best.

- To place a cardioid microphone, cup your hand behind your ear and listen. Move around the player or sound source until you find a spot that sounds best.

- For a stereo pair, cup your hands behind both ears. Move around the player or sound source until you find a spot that sounds best.

Before you start swapping gear, know that the three most important factors in getting the sound you want are mic position, mic position, and mic position. Get the instrument to make the sound you want to record first, then use the cover-your-ears technique to find the sweet spot, position the mic, then listen. Remember that if you can't hear it, you can't record it. Don't be afraid to repeat as much as necessary or to experiment if you're not getting the results you want.

General Placement Techniques

The following are some general issues and techniques to consider before placing a mic:

- One of the reasons for close-miking is to avoid leakage into other mics, which means that the engineer can have more flexibility later in balancing the ensemble in the mix. That said, give the mic as much distance from the source as possible in order to let the sound develop and be captured naturally.

- Mics can't effectively be placed by sight until you have experience with the player, the room you're recording in, the mics you're using, and the signal path. If at least one of these elements is unknown, at least some experimentation is in order until the best placement is found. It's okay to start from a place that you know has worked in the past, but be prepared to experiment with the placement a bit since each instrument and situation is different.

- If the reflections of the room are important to the final sound, start with any mics that are used to pick up the room first, then add the mics that act as support to the room mics.

- From 200Hz to 600Hz is where the proximity effect often shows up and is one reason why many engineers continually cut EQ in this range. If many directional microphones are being used in a close fashion, they will all be subject to proximity effect, and you should expect a buildup of this frequency range in the mix as a result.

- One way to capture a larger-than-life sound is by recording a sound that is not as loud as the level that the recording will most likely be played back at. For electric guitars, for instance, sometimes a small 5-watt amp into an 8-inch speaker can sound larger than a cranked full Marshall stack.

AVOIDING PHASE CANCELLATION

One thing you'll see frequently throughout this book is references to a phenomenon known as phase cancellation. One of the most overlooked aspects of recording is how microphones interact with each other, both acoustically and electrically.

This is a concern because it only takes a single out-of-phase mic to destroy the sound of a multi-miked instrument like a drum kit, and if not corrected, the sound may never be able to be fixed.

Without getting into a long-winded technical explanation, microphone phase refers to the fact that the output from all microphones used on the session should be pushing and pulling together as one. If one mic is pushing while another is pulling, they cancel each other out at certain frequencies. In Figure 5.1, when mic #1's signal peaks, mic #2's signal valleys. They cancel each other out at that frequency, and the result is a very weak-sounding signal when mixed together.

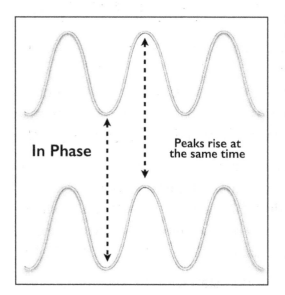

Figure 5.1: Two microphones out of phase
© 2023 Bobby Owsinski (Source: iStock Photo)

In Figure 5.2, both mics are pushing and pulling together. Their signal peaks happen at the same time, as do their valleys. As a result, their signals reinforce one another.

Figure 5.2: Two microphones in phase
© 2023 Bobby Owsinski (Source: iStock Photo)

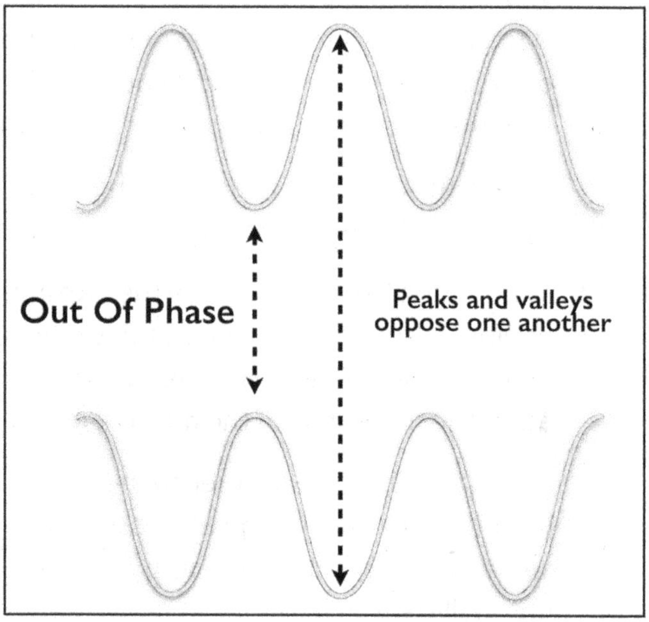

Naturally, all frequencies won't either reinforce or cancel. The idea is to have as few cancel as possible.

Electronic Phase Cancellation

There are two types of phase cancellation: electronic and acoustic. Most of the phase problems we'll discuss in the book are acoustic in nature, but there's also an instance of electronic phase cancellation that you should know about as well. Keep in mind that this has nothing to do with mic placement, as it's strictly an electronic problem that never shows up until multiple mics are used.

Electronic phase problems are almost always caused by a cable in the studio (usually a mic cable) that's been mis-wired during an install, repaired incorrectly, or originally wired incorrectly from the factory (which is rare). The polarity check is used mainly to be sure that all mics are pushing and pulling the same way at the same time, and to check for mis-wired cables, which happens more than you'd think even in some of the best facilities in the world.

Checking The Polarity

Checking microphone polarity should be one of the first things that an engineer does after all the mics are wired up and tested, especially if you're working in an unfamiliar studio. This is especially true before a tracking session where a lot of mics will be used, since having just one mic out of phase can cause an uncorrectable sonic problem that will most likely haunt the recording forever. A session that is in-phase will sound bigger and punchier, while just one out-of-phase mic can make the entire mix sound tiny and weak.

For this test we'll be using the phase switch on the mic preamp, DAW interface, or console, since it's really a polarity switch that changes the phase by 180 degrees at all frequencies by swapping pins 2 and 3 of a balanced microphone line. Here's the test:

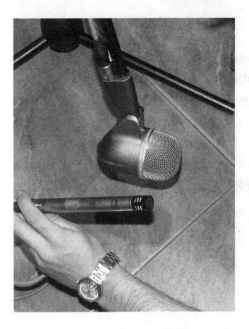

1. After the mics are set up, wired, and checked but not necessarily placed, pick one mic that can be easily moved. This can be a scratch vocal mic, a hat mic, or a guitar mic; it doesn't matter as long as it works, sounds good to begin with (meaning that it's not defective), and can move next to the farthest mic used in the session. This mic will become our "gold standard."

2. With the gold-standard mic in hand, move it next to the kick drum mic (or any other mic that you wish to test, for that matter). Place both mics together so the capsules touch and speak into them from about a foot away (the distance isn't critical—see Figure 5.3).

Figure 5.3: Checking polarity
© 2023 Bobby Owsinski

3. Bring up the faders on both mics so the audio level (not the fader position) is equal on both. You can check this by making sure that the level meters read the same.

4. Flip the phase of the mic under testing (in this case, the kick mic) and choose the position that gives you the most low end.

5. Repeat with all the other mics used on the session.

Remember, you're not flipping the phase of the gold-standard mic, only the one that you're testing.

Acoustic Phase Cancellation

An acoustic phasing problem occurs when two mics are too close to one another and pick up the sound from the same instrument, only one is picking it up a little later in time than the first because it's a little farther away (see Figure 5.4). The phase check will make sure that you minimize the interference between the mics when they're placed.

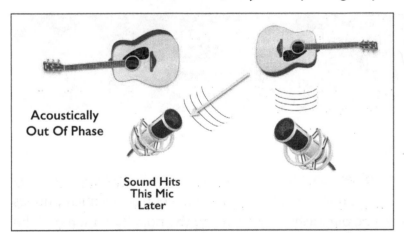

Figure 5.4: Acoustically out of phase
© 2023 Bobby Owsinski (Source: iStock Photo)

With acoustic phase problems, the sounds cancel out only at certain frequencies, which usually makes the sound of the two mixed together either hollow-sounding or just lacking in depth and bottom end. The way to eliminate the problem is by moving mic #2 a little farther away from mic #1, or if the mics are directional, making sure that each one is pointing directly at the source that they're trying to capture.

93 | Microphone Placement Fundamentals

The 3-to-1 Principle

The 3-to-1 principle is an important consideration when it comes to any multi-mic setup because, if you observe the rule, you can nearly eliminate any phase problems before they start. The 3-to-1 principle states that in order to avoid phase cancellation between two microphones, a second mic should never be within three times the distance that the first mic is from its source. That means that if a pair of microphones is placed in front of a piano at a distance of 1 foot away, the separation between the two mics should be at least 3 feet in order to avoid any cancellation. If the distance from the source is 2 feet, the distance between mics should be at least 6 feet (see Figure 5.5).

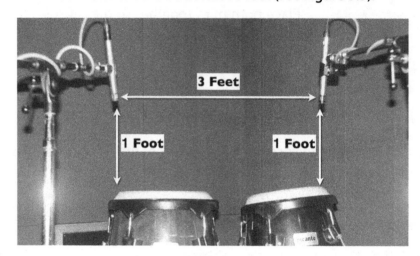

Figure 5.5: The 3-to-1 principle
© 2023 Bobby Owsinski

This principle is not a hard-and-fast rule and is frequently violated when it comes to real-life recording, but it's certainly a good guideline for eliminating any phase problems that might arise. Remember, if you record something with a phase problem, no amount of EQ or processing afterward can ever make it right.

MIC PLACEMENTS MOST LIKELY TO CAUSE PHASE PROBLEMS
- Mics that are facing each other (such as on the opposite sides of a drum)
- Mics that are facing the floor (just angle them a bit)

Checking The Phase By Listening

As said previously, the chances for a phase problem are far greater on the drum kit because it usually has more mics on it than any other instrument. Understand that you will never have all microphones completely in phase, but some problems will be diminished by reversing the polarity on some of the channels. The only way to determine this is through experimentation and listening.

For this check we'll again use the phase switch on the console, mic preamp, or DAW. Flipping the phase switch may cause the problem frequencies to come closer in phase, or it may put them farther away, or it may have no effect at all.

1. Record a sample of the drummer playing, then listen to the overheads panned in stereo, then listen to them in mono by panning them both to the center. Flip the phase switch on only one of the overheads. If the low frequencies seem to drop out, flip the phase switch back to the original position. If it sounds like there's more low end, leave it in this position. After you've found the correct setting, pan the channels back into stereo and go on to the next step.

2. If the overheads still sound thin or swishy and you know that their polarity is correct (they're pushing and pulling at the same time), then place them in a different position, perhaps using them as a stereo pair (see Chapter 6, "Basic Stereo Techniques," for more on stereo miking) or placed farther apart.

3. Once you're pleased with the sound coming from the overheads, add the kick drum. Switch the polarity on the kick while listening to the overheads and stay with the position that has the fullest sound.

4. Bring up the snare mic. Press the phase button on the console or preamp. Does it sound better inverted or not? Now place the entire mix in mono (if possible) and see whether it still sounds better.

5. Keep following this procedure for each microphone. On each one, listen to how the mic sits in the mix, then listen to it with the phase inverted, and then do the same thing in mono. In each case, use the phase switch position that gives you the fullest sound with the most low end.

If you have two kick mics, check the phase of the inside kick mic against the overheads and then the outside kick against the inside mic. Sometimes you might need to move the outside mic to find a more phase-compatible position.

Ultimately, you cannot avoid phase cancellation, you can only make sure it sounds as good as possible.

TIP: Remember that one position of the phase switch may sound fuller than the other, and that's the one to choose. If neither position seems to make a difference, choose the position you started with.

Checking Phase With A Phase Meter

Many traditional hardware consoles have a built-in phase meter, and now most DAWs have a similar phase meter plug-in. Being able to properly read it can go a long way in determining whether you have any phase problems.

If you feed the stereo buss a mono signal on both channels and it's in phase, the needle should read all the way to the right, or + (see Figure 5.6). If the level of both channels is identical but out of phase, the needle will read all the way to the left, or -. This is the condition that you're trying to avoid, since if you sum the two channels together they'll cancel out in mono. If there's no relationship between the two channels, the needle will sit in the middle. For most stereo program material that has a lot of identical content on both channels, it will wander around mostly in the + side of the scale and occasionally to the - side (which is okay). If it stays mostly on the minus side of the scale, then something is drastically out of phase and requires that you sweep through the channels again to find the channel or channels that are out of phase, or that you move the mics.

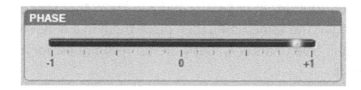

Figure 5.6: A phase meter with an in-phase signal
© 2023 Bobby Owsinski (Source: Avid)

Checking Phase With An Oscilloscope Plugin

Another way to check your phase is to look at an oscilloscope plugin. Remember that the nature of music means that there will always be some phase discrepancies between any two mics picking up the same sound from different locations, so you will never see perfect absolute phase alignment, but you still should be able to get it easily in the ballpark.

1. Insert an oscilloscope plug-in, such as a DigiRack Phase Scope or T-RackS Meter across the stereo buss in the DAW.

2. Send the same mono signal to both stereo channels. You should see a straight, diagonal line slanting to the right (see Figure 5.7).

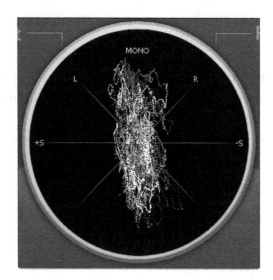

Figure 5.7: An in-phase scope signal
© 2023 Bobby Owsinski (Source: Avid, Flux:: Sound And Picture Development)

3. Now, play your stereo mix, and you'll see "scrambled eggs" as the line opens up into an irregular and constantly changing circle. The more "open" the circle, the greater the phase difference between the mics. What you want to avoid is a condition where the circle starts slanting to the left (see Figure 5.8).

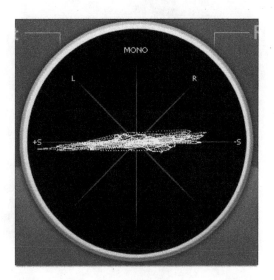

Figure 5.8: An out-of-phase scope signal
© 2023 Bobby Owsinski (Source: Avid, Flux:: Sound And Picture Development)

TIP: If something doesn't sound right, there are a lot of things to change before you reach for the EQ. Try the following in this order:

1. *Change the source, if possible (the instrument you are miking).*
2. *Change the mic placement.*
3. *Change the placement in the room.*
4. *Change the mic.*
5. *Change the mic preamplifier.*
6. *Change the amount of compression and/or limiting (from none to a lot).*
7. *Change the room (the actual room you are recording in).*
8. *Change the player.*
9. *Come back and try it another day.*

Although numbers 7 through 9 may not be your call, they're worth suggesting to the producer.

97 | Microphone Placement Fundamentals

BASIC STEREO TECHNIQUES

Even if you never intend to record an ensemble larger than a standard rock, pop, or jazz rhythm section, a good grasp of the many stereo recording techniques is essential and will come in handy sooner or later. Stereo miking is commonly used when recording drum kits, pianos, string sections, Leslie speakers, and the like, and can certainly be applied to just about any recording situation.

While we won't discuss these techniques in an orchestral music sense (where a lot of knowledge beyond the scope of this book is a necessity), here's a basic overview of the many methods of stereo miking.

First of all, stereo miking is an improvement over mono miking because it provides:

- A sense of the soundfield from left to right
- A sense of depth or distance between each instrument
- A sense of distance of the ensemble from the listener
- A spatial sense of the acoustic environment: the ambience or hall reverberation

GENERAL TYPES OF STEREO MIKING

There are four general mic techniques used for stereo recording, each with a different sound and different sets of benefits and disadvantages:

- Coincident pair (including X/Y, M-S, and Blumlein)
- Spaced pair
- Near-coincident pair (the famous ORTF method)
- Baffled-omni pair or artificial head

COINCIDENT PAIR

A coincident pair consists of two directional mics mounted so that their grilles are nearly touching, but with their diaphragms angled apart in such a way that they aim approximately toward the left and right sides of the ensemble or instrument. For example, two cardioid microphones can be mounted angled apart, their grilles one above the other. The greater the angle between microphones, the wider the stereo spread.

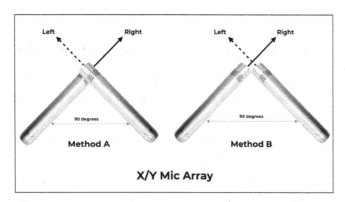

Figure 6.1: Two different X/Y methods
© 2023 Bobby Owsinski

X/Y

While there are several variations of the coincident pair, the X/Y configuration is the easiest and mostly widely used. X/Y requires two identical directional microphones. Unlike what you may think, the mics shells are not crossed in an X pattern in this configuration. In fact, it's the mic capsules that are placed as close as possible to one another in a 90-degree angle (see Figure 6.1).

M-S

M-S stands for mid-side and consists again of two microphones: a directional mic (an omni can be substituted as well) pointed toward the sound source and a figure-8 mic pointed toward the sides. Once again, the mics are positioned so that their capsules are as close to touching as possible (see Figure 6.2).

Figure 6.2: M-S miking
© 2023 Bobby Owsinski (Source: iStock Photo)

M-S is great for stereo imaging, especially when most of the sound is coming from the center of the ensemble. Because of this, this technique can be less effective on large groups, favoring the middle voices that the mics are closest to.

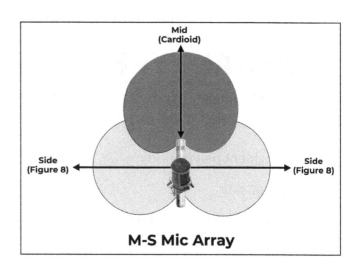

M-S doesn't have many phase problems in stereo and has excellent mono compatibility, which can make it the best way to record room and ambience under the right circumstances.

In many cases it can sound more natural than a spaced pair, which is covered later in the chapter. If the source is extra large, sometimes using M-S alone will require too much distance away from the ensemble to get the whole section or choir into perspective, so multiple mic locations must be used.

If a narrower pickup pattern is required to attenuate the hall sound, then a directional mic such as a cardioid or even a hypercardioid will work for the "M" mic. Just be aware that you may be sacrificing low-end response as a result.

For best placement, walk around the room and listen to where the instrument or sound source sounds best. Note the balance of instrument to room, and the stereo image of the room as well. When you have found a location, set up the directional mic where the middle of your head was.

M-S Decoding

Listening to either of these mics alone may sound okay or may even sound horribly bad. That's because in order to make this system work, the mic's output signals need an additional decoding step to reproduce a faithful stereo image. The directional mic creates a positive voltage from any signal it captures, and the bidirectional mic creates a positive voltage from anything coming from the left and a negative voltage from anything coming from the right. As a result, you need to decode the two signals to create the proper stereo effect.

While you can buy an M-S decoder, you can easily emulate one with three channels on your console or DAW. On one channel, bring up the cardioid (M) forward-facing mic. Copy the figure-8 mic (S) to two additional channels in your DAW. Pan both channels to one side (such as hard left) and then flip the phase of the second "S" channel and bring up the level until the two channels cancel 100 percent.

Now pan the first "S" channel hard left and the second "S" channel hard right, balance the cardioid (M) channel with your pair of "S" channels, and you have your M-S decode matrix.

A nice additional feature of this method is that you can vary the amount of room sound (or change the "focus") by varying the level of the bidirectional "S" mic.

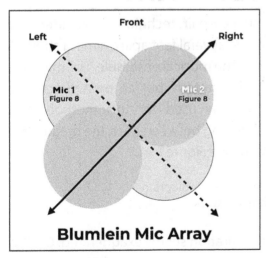

Figure 6.3: A Blumlein array
© 2023 Bobby Owsinski

Blumlein Array

Developed for EMI in 1935 by famed audio pioneer Alan Blumlein, the Blumlein stereo setup is a coincident stereo technique that uses two bidirectional microphones positioned close to one another, then angled at 90 degrees to each other. This stereo technique will normally give the best results when used at closer distances to the sound source, since the lower frequencies tend to roll off at larger distances away from the sound source. The Blumlein stereo technique has greater channel separation than X/Y stereo, but it has the disadvantage that sound sources located behind the stereo pair also will be picked up and even be reproduced with inverted phase (see Figure 6.3).

The Stereo Microphone

Although not normally thought of as a coincident mic pair, a stereo mic uses two coincident mic capsules mounted in a single housing for convenience (see Figure 6.4). Because of their close proximity to one another, this method provides the easiest coincident mic setup, since accessories such as a stereo bar (see Figure 6.5) or multiple mic stands are not required.

Figure 6.4: Royer SF-12 stereo microphone
Courtesy of Royer Labs

Figure 6.5: A stereo bar
© 2023 Bobby Owsinski

COINCIDENT-PAIR FEATURES:
- Imaging is very good.
- Stereo spread ranges from narrow to accurate.
- Signals are mono-compatible.

SPACED PAIR

With the spaced-pair technique, two identical mics are placed several feet apart, aiming straight ahead toward the musical ensemble. The mics can have any polar pattern, but an omnidirectional pattern is usually used for this method. The greater the spacing between mics, the greater the stereo spread (see Figure 6.6).

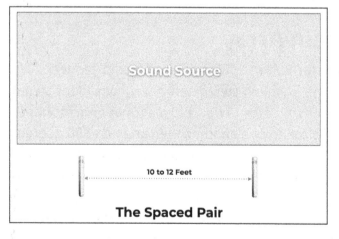

Figure 6.6: Spaced-pair diagram
© 2023 Bobby Owsinski

If the spacing between mics is too far apart though, the stereo separation seems exaggerated. On the other hand, if the mics are too close together, there will be an inadequate stereo spread. In an attempt to obtain a good musical balance, the mics are usually placed about 10 or 12 feet apart, but such spacing may result in exaggerated separation. One solution is to place a third microphone midway between the original pair and mix its output to both channels. That way, the ensemble is recorded with

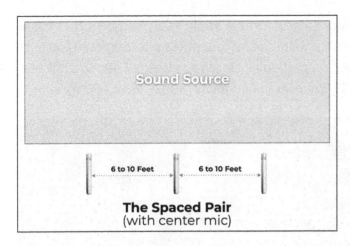

a good balance, and the stereo spread is not as exaggerated (see Figure 6.7).

Figure 6.7: Spaced pair with a center mic
© 2023 Bobby Owsinski

The spaced-pair method tends to make off-center images relatively unfocused or hard to localize. In addition, combining both mics to mono sometimes causes phase cancellations at various frequencies.

The advantage with spaced miking is a warm sense of ambience in which room reflections or concert-hall reverberation seems to surround the instruments and sometimes the listener. Another advantage of the spaced-mic technique is the ability to use omnidirectional microphones. An omni condenser mic sometimes can have a better low-frequency response than a unidirectional condenser mic and tends to have a smoother response and less off-axis coloration.

The Decca Tree

A variation of the spaced pair is the Decca Tree, which is essentially a spaced pair with a center mic connected to a custom stand and suspended over the conductor during orchestral recording (see Figure 6.8). Decca Records, who had a long tradition of developing experimental recording techniques including surround sound and proprietary recording equipment, developed the Decca Tree in 1950s as compromise between the purist stereo pair and multi-mic arrays that were being used. Apart from individual engineer's choice of mic, it remains largely unchanged to this day. It's still in use in film scoring, classical orchestral, and opera recording, as it produces a very spacious stereo image with good localization.

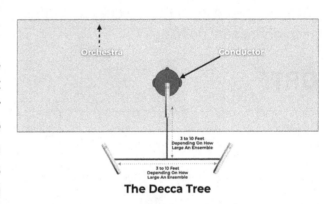

Figure 6.8: The Decca Tree
© 2023 Bobby Owsinski

The classic Decca Tree setup uses three Neumann M 50s arranged in a triangle 10 to 12 feet above the conductor's position, although the spacing varies with venue and size of ensemble. The name comes from engineer Arthur Haddy, who declared, "It looks like a bloody Christmas tree!"

The left mic is panned left, the right mic is panned right, and the center mic is panned to the center with additional mics (called spot mics or sweeteners) used over the violins (usually panned to the left), cellos (usually panned to the right), harp and timpani, and the soloist.

The distance between mics depends on the size of the ensemble. For an orchestra, the left and right mics are 8 to 10 feet apart, with the center about 6 to 7 feet in front of the left-right axis.

The Neumann M 50 was found to be the perfect microphone for this application as it becomes more directional above 1 kHz, and has an extended bass and high frequency boost. It's a little-known fact that Decca has used mics other than the M 50 on the tree. In particular, Decca has used M 49s and KM 56s, but modern substitutions include the TLM 50, M 150, Brauner VM-1, or DPA 4003's with APE spheres.

> **SPACED-PAIR FEATURES:**
> - Off-center images are diffuse.
> - Stereo spread tends to be exaggerated unless a third center mic is used.
> - Provides a warm sense of ambience.
> - Phasing problems are possible.

NEAR-COINCIDENT PAIR

The near-coincident stereo miking technique is similar to the coincident pair configurations except the mics are set up with some distance between the microphone capsules instead of next to one another. There are two different techniques that differ only by the distance and the angle at which the two microphones are pointing.

ORTF

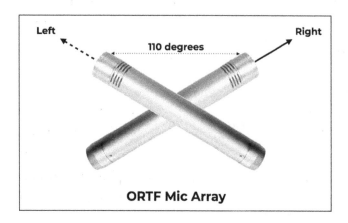

Figure 6.9: An ORTF stereo setup
© 2023 Bobby Owsinski

The most common example of the near-coincident method is the ORTF system, which uses two cardioids angled 110 degrees apart and spaced 7 inches (17 cm) away from each other horizontally (ORTF stands for Office de Radiodifusion Television Française or the Office of French Radio and Television Broadcasting). The spacing is the distance between each ear of a typical human head. This method tends to provide accurate localization; that is, the instruments appear in the soundfield of the recording as they're placed in the room or on stage. ORTF provides a much greater sense of space due to time/phase differences, since the capsules are as far apart as your ears (see Figure 6.9).

NOS

NOS, which stands for Nederlandshe Omroep Stichting or the Netherlands Broadcasting System, places two cardioid microphones 11.8 inches apart (30 cm) at a 90-degree angle from one another (see Figure 6.10). NOS has better mono compatibility than ORTF and a stronger center image as well. It's also somewhat easier to set up than ORFT because the angle is easier to measure.

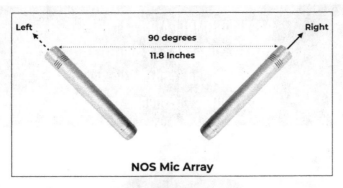

Figure 6.10: An NOS stereo setup
© 2023 Bobby Owsinski

> **NEAR-COINCIDENT-PAIR FEATURES:**
> - Sharp imaging
> - Accurate stereo spread
> - A greater sense of air and depth than coincident methods

BAFFLED-OMNI PAIR

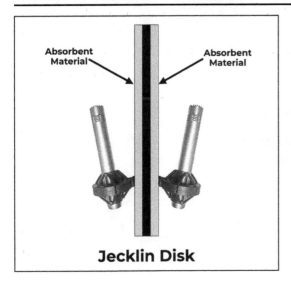

A baffled-omni pair tries to emulate the way our ears are placed on our heads and, therefore, the way we hear.

In this method, two omnidirectional mics are separated a few inches by a baffle between them. The baffle can be any hard surface covered with absorbent foam (as in the Jecklin Disk - see Figure 6.11). Another configuration uses a hard sphere for a baffle with the mics flush-mounted on opposite sides, as in the Schoeps spherical mic (see Figure 6.12).

Figure 6.11: Jecklin Disk
© 2023 Bobby Owsinski

Figure 6.12: Schoeps KFM 360 spherical mic
Courtesy of Schoeps GmbH

With the baffled-omni pair, the level, time, and spectral differences between channels create the stereo image. The omni condenser mics used in this method provide excellent low-frequency response. Also falling into this category is the dummy head, such as the Neumann KU 100 (see Figure 6.13).

BAFFLED-OMNI-PAIR FEATURES:
- Images are sharp
- Stereo spread tends to be accurate
- Low-frequency response is excellent

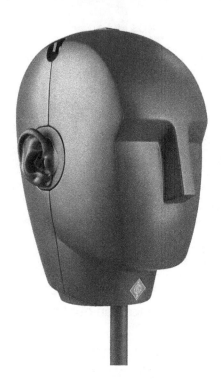

Figure 6.13: Neumann KU 100 dummy head
Courtesy of Neumann USA

PREPARING THE DRUM KIT FOR RECORDING

Probably the single most troublesome instrument to record is the drum kit. Engineers obsess over the drum sound, and well they should, since the drums are the heartbeat of virtually all modern music. It's a fact that drums that sound small in the track usually will make the rest of the track sound small as well, regardless of how well everything else is recorded.

Unfortunately, most inexperienced engineers attribute their drum sound to their recording gear when in reality that's not the case at all. When it comes right down to it, the main reason for a great drum sound comes from the drum kit itself and the talent of the drummer. Get a great-sounding kit and a killer drummer, and it's difficult to make a bad sounding drum recording. On the other hand, a bad sounding kit with a wimpy drummer isn't going to record well regardless of how much high-end recording gear you throw at it.

Unless a drummer is used to playing in the studio, there's a good chance that his or her drum kit won't record well, and that occurs for any number of reasons. It could be because of old, beat-up heads (the worst offender), bad tuning, uneven bearing edges of the drum shells, or a piece of drum hardware is defective. Whatever it might be, drums that might be adequate or even sound pretty good in a live situation don't always cut it when put under the scrutiny of the recording studio.

While many engineers are willing to spend whatever time it takes to make the drums sound as good as possible, most just don't have the know-how or the time to improve the sound of the set before it gets under the mics. As a result, virtually all big-budget projects either rent a kit specifically for recording or hire a drum tuner, because no matter how great your signal chain is, if the drum sound in the room isn't up to par, there's not much the engineer can do to help.

Because this book is about getting great sounds, it's important to make the drums sound their best before you even place a mic. But before we go there, just what constitutes a great-sounding drum kit?

THE KEYS TO A GREAT SOUNDING DRUM KIT

It's true that different people have different ideas of what constitutes a great sounding kit, and that's different for different musical genres, but in the studio it usually means a kit that's well-tuned and free of buzzes and sympathetic vibrations. Free of sympathetic vibrations means that when you hit the snare drum, for instance, the toms don't ring along with it. Or if you hit the rack toms, the snare and the other toms don't ring along as well. The way to achieve this is all in the tuning and the kit maintenance, which we'll check out in depth later in the chapter, but first let's learn a little bit about drums themselves, since it helps to have a basic idea of why they sound the way they do.

Drum Construction

Here are the things that affect the sound of a drum:

- **Shell size** has the most impact on the natural pitch of a drum. The larger the diameter, the lower the natural pitch, although you can obviously change this a bit by tuning the heads.
- **Shell depth** is mostly responsible for how loud the drum will be and, to some degree, the articulation of the sound. This means that a shallow shell (say, a 9" tom) doesn't have as much surface area as a larger one, so the sound doesn't ring as long and has a sharper attack.
- **Shell thickness** is usually overlooked as a contributing factor to the sound of a drum. Thinner shells actually are more resonant since they're easier to excite because they have a lower mass than a heavier, thicker shell.
- **Shell material** used to make the drum shell is the most responsible for the tone of the drums. Here are the most commonly used drum-shell materials.
  - » Maple is the most prized construction material by drummers, primarily because the sound is so even across the drum frequency spectrum.
  - » Mahogany sounds warmer than maple since the low end is increased.
  - » Birch is very hard and dense, which results in a brighter drum with a lot less low end than maple.
  - » Poplar has a sound very similar to birch, with a bright top end and less bottom.
  - » Basswood exhibits an increased low end that's similar to mahogany.
  - » Luaan has a warmer sound with less top end, similar to mahogany.
- **Shell interior** has a lot to do with the pitch of the drum. A rough interior produces a less resonant drum, since the roughness breaks up the interior reflections. A smooth interior results in a more resonant drum, which means it's easier to tune and control.

- **Bearing edges** means the cut at the edge of a drum shell where the hoops are attached. The way the bearing edge is cut can affect not only the pitch of the drum, but also how well it tunes. The sharper the cut, the brighter the drum.
- **Hoop type and the number of lugs** used to seat the drum heads determines how the drum will sound as well. In general, the thicker the hoop, the easier the drum will be to tune. Fewer lugs provide more complex overtones. Stamped hoops get a warmer tone than from die-cast hoops. Aluminum gives a high pitch, while brass provides more overtones. Die-cast hoops are generally both thicker and stronger than stamped hoops, so the drum becomes easier to tune. There are fewer overtones as a byproduct. Wood hoops come in different thicknesses, so they can be made to sound like either a stamped or a cast hoop, only brighter.

Drum Heads

If there's one simple action that you can take to improve the tone of the drums, it's to replace the old heads with fresh new ones. Even drums that normally sound pretty good will sound wimpy and dead when played with old heads that have dings and dents in them.

Of course, all heads are not created equal, so here's a simple chart to help with the selection, as recommended by Ross Garfield, the Drum Doctor.

Table 7.1: Drum Head Recommendations

DRUM	HEAD TYPE
Kick	Start with a head like a Remo Powerstroke 3 and add a damping ring to eliminate the higher frequencies. For more punch, go to a heavier head with more attack and less sustain, such as a Remo Emperor.
Snare	Start with a head with a lot of sustain and tone, such as a white Ambassador, or a dampened head, such as a coated black dot Ambassador, on the snare top, and either a thin head, such as a clear Diplomat, or a coated Ambassador on the bottom. If 6-1/2" or deeper, go to a Diplomat. If too much ring, go to a heavier coated Remo Emperor. If not enough ring, go to a lighter Remo Ambassador or Diplomat.
Toms	Start with white Remo Ambassadors on top and thin clear Ambassadors on the bottom.

TIP: The single biggest improvement to the tone of a drum is a set of fresh heads.

The Drum Tuning Technique

Unless you're already a drummer, you probably aren't aware of the proper way to tune a drum. The process is actually quite simple, but it does take some time and experimentation. The idea is to make sure that all of the tension rods that hold the head on have the same tension at each lug. What you want is for the pitch to sound the same at each lug as you tap near it. Here's how it's done:

1. Hit the head an inch in front of each lug of the drum. Is the sound the same at each lug?

2. Using a drum key, adjust the tension so that the sound is the same at each lug. Is the sound the same at each lug now?

3. When the pitch (the tension) is the same at each lug, then hit the drum in the center. It should have a nice, even decay.

4. Using the same technique, tune the bottom head to the same pitch as the top head. What does the drum sound like now when you hit it in the center? Is the tone even? Is the decay even? Are there any overtones?

***TIP:** For faster and more even tuning, adjust the lugs in a crisscross pattern, as shown in Figure 7.1.*

Figure 7.1: The crisscross drum-tuning pattern
© 2023 Bobby Owsinski

While it's important to learn how to tune a drum by ear, there are tools to make it a bit easier. The DrumDial measures the tension at each lug while the Evans Torque Key measures the drum lug torque for repeatable tuning.

Tuning between the Top and Bottom Heads

There are three ways to tune drums that use a top and bottom head:

- **The top and bottom heads are tuned to the same pitch.** This provides the purest tone and the longest sustain.
- **The bottom head is tuned lower than the top.** This provides a deep sound with a lot of sustain as well as a pitch drop or "growl."
- **The bottom is tuned higher (tighter) than the top.** This also produces a pitch-drop sound, although it's a bit shallower and has a shorter sustain.

When the drum is first tuned, both heads are tuned to the same pitch. After the correct pitch for the top head is selected, tune the bottom head anywhere from a pitch interval of a third to a fifth away from the top head, if that's the sound you're looking for.

Tuning Between Drums

Many drummers prefer to tune drums a fourth away (C to F on a piano, for instance) from each other, while others will tune drums as far away from each other as they can. The limiting factor is how low you can tune the floor tom before it begins to sound bad or too much like a kick drum.

1. Start with your smallest tom and tune the top head close to the pitch where it resonates well, being careful not to tune it too low.

2. To tune in fourths, go to the next largest drum and sing the "Wedding March," or "Here comes the bride." The interval between "Here" and "comes" is a fourth, so "comes" would be the tuning pitch of the high tom, and "Here" would be the pitch of the lower.

3. Continue tuning the rest of the toms using the same method.

TIP: When drums are close to the same diameter, tune the smaller one higher and the larger one lower.

Tuning Tips

Here's a list of tips for the most common problems found with each drum of a typical kit. For additional tips and tricks, read the upcoming "Interview with 'The Drum Doctor' Ross Garfield," or check out *The Drum Recording Handbook* 2nd edition (Hal Leonard).

If the snares buzz when the toms are hit:

- Check that the snares are straight.
- Check to see whether the snares are flat and centered on the drum.
- Loosen the bottom head.

- Retune the offending toms.
- Use an alternate snare drum.

If the snare has too much ring:

- Tune the heads lower.
- Use a heavier head, such as a coated Remo Emperor.
- Use a full or partial muffling ring, or add some tape or Moongel.

If the kick drum isn't punchy and lacks power in the context of the music:

- Try increasing and decreasing the amount of muffling in the drum. Try anything from a sandbag to a different blanket or pillow.
- Change to a heavier, uncoated head, such as a clear Emperor or Powerstroke 3.
- Change to a thinner front head or one with a larger cutout.

If one or more of the toms are difficult to tune or have an unwanted "growl":

- Check the top heads for dents and replace as necessary.
- Check the evenness of tension all around on the top and bottom heads.
- Tighten the bottom head.

Cymbals

Most drummers who play live more than in the studio usually have heavier cymbals because they need to be loud enough to cut through the rest of the band on stage. Plus heavier cymbals also last longer. The problem is that heavier cymbals can have a gong-like quality when recording, so lighter cymbals are usually preferred.

Be careful when you mix different cymbal weights when you're recording because of the different volumes that each will have. For example, if you use a couple of thick cymbals, a thinner cymbal might disappear in the mix because it's not as loud.

INTERVIEW WITH "THE DRUM DOCTOR" ROSS GARFIELD

Anyone recording in Los Angeles certainly knows about the Drum Doctors, *the* place in town to either rent a great-sounding kit or have your kit fine-tuned. Ross Garfield is the "Drum Doctor," and his knowledge of what it takes to make drums sound great under the microphones may be unlike any other on the planet. Having made the drums sound great on platinum-selling recordings for the likes of The Rolling Stones, Paul McCartney, Bruce Springsteen, Rod Stewart, Metallica, Kendrick Lamar, Beyonce, Red Hot Chili Peppers, Foo Fighters, Lenny Kravitz, Michael Jackson, and many more than what can comfortably fit on this page, Ross agreed to share his insights on drum tuning.

What's the one thing that you find wrong with most drum kits that you run into?

I think most guys don't know how to tune their drums, to be blunt. I can usually take even a cheap starter set and get it sounding good under the microphones if I have the time. It's really a matter of people getting in there and changing their heads a lot. Not for the fact of putting fresh heads on as much as the fact that they're taking their drums apart and putting them back together and tuning them each time. The repetition is a big part of it. Most people are afraid to take the heads off their drums.

When I get called into a session that can't afford to use my drums and they just want me to tune theirs, the first thing I'll do is put a fresh set of heads on.

How long does it take you to tune a set that needs some help?

Usually well under an hour. If I have to change all the heads and tune them up, it'll take about an hour before we can start listening through the mics. I try to tune them to what I think they should be, then when we open up the mics and hear all the little things magnified, I'll modify it. Once the drummer starts playing, I like to go into the control room and listen to how they sound when they're played. Then once the band starts, I'll see how the drum sound fits with the other instruments.

What makes a drum kit sound great?

I always look for a richness in tone. Even when a snare drum is tuned high, I look for that richness. For example, on a snare drum I like the ring of the drum to last and decay with the snares. I don't like the ring to go past the snares. And I like the toms to have a nice, even decay. Usually I'll tune the drums so that the smallest drums have a shorter decay and the decay gets longer as the drums get bigger. I think that's pleasing.

What's the next step to making drums sound good after you change the heads?

I tune the drums on the high side for starters. For tuning, you've got to keep all of the tension rods even so they have the same tension at each lug. You hit the head an inch in front of the lug, and if you do it enough times you'll hear which ones are higher and which are lower. The pitch should be the same at each lug; then when you hit it in the center, you should have a nice, even decay. I do that at the top and the bottom head.

Are they both tuned to the same pitch?

I start it that way and then take the bottom head down a third to a fifth below the top head.

I've been in awe of the way you can get each drum to sound so separate without any sympathetic vibrations from the other drums. Even when the other drums do vibrate, it's still pleasing. How do you do that?

Part of that is having good drums, and that's the reason why I have so many, so I can cherry-pick the ones that sound really good together. The other thing is to have the edges of the shells cut properly. If you take the heads off, the edges should be flat. I check it with a piece of granite that I had cut that's perfectly flat and about 2 inches thick. I'll put the shell on the granite and have a light over the top of the shell. Then I'll get down at where the edge of the drum hits the granite. If you see light at any point, then you have a low spot. So that's the first thing: to make sure that your drums are "true."

The edges should be looked at anyway, because you don't want to have a flat drum with a square edge; you want it to have a bevel to it. If you have a problem with a drum, you should just send it in to the manufacturer. I don't recommend anyone trying to cut the edges of their drums themselves. It doesn't cost that much, and it's something that should be looked at by someone who knows what to look for.

Once you get those factors in play, then tuning is a lot easier. I tend to tune each drum as far apart as the song will permit. It's easy to get the right spread between a 13- and a 16-inch tom, but it's more difficult to get it between a 12 and a 13. What I try to do is to take the 12 up to a higher register and the 13 down a little. The trick to all that is the snare drum because the biggest problem that people have is when they hit the snare drum, there's a sympathetic vibration with the toms.

The way I look at that is to get the snare drum where you want it first, because it's way more important than the way the toms are tuned. You hear that snare on at least every two and four. The kick and snare are the two most important drums, and I tune the toms around that and make sure that the rack toms aren't being set off by the snare. The snare is probably the most important drum in the set because for me it's the voice of the song. I try to pick the right snare drum for the song because that's where you get the character.

Do you tune to the key of a song?

Not intentionally. I have people who ask me to do that, and I will if that's what they want, but usually I just tune it so it sounds good with the key of the song. If there's a ring in the snare, I try to get it to ring in the key of the song, but sometimes I want the kit just to stand on its own because if it is tuned in the key of the song and one of the players hits the note that the snare or kick is tuned to, then the drum kind of gets covered up, so I tend to make it sound good with the song rather than in-pitch with the song.

Would you tune things differently if you have a heavy hitter as opposed to someone with a light touch?

Yeah, a heavy hitter will get more low end out of a drum that's tuned higher just because of the way he hits, so I usually tune a drum a little tighter. I might move into different heads as well, like an Emperor or something thicker.

How about the kick drum? It's the drum that engineers spend the most time on.

It's weird for me because I always find them to be pretty easy because you muffle the kick drum on almost every session, and when you do it makes tuning easier. On the other hand, a tom has as much life as possible with no muffling.

What I would recommend is to take a down pillow and set it up so that it's sitting inside the drum, touching both heads. From there you can experiment, so if you want a deader, drier sound, then you push more pillow against the batter head; and if you want it livelier, then you push it against the front head. That's one way to go.

Another way to go is to take three or four bath towels and fold one of them so it's touching both heads. If that's not enough, then put another one in against both heads on top of the first one. If that's not enough, then put another one in. Just fold them neatly so that they're touching both heads. That's a good place to start; then experiment from there.

Do you prefer a hole in the front head?

It makes it easier. I do some things without holes in the front head, but having it really makes it easy to adjust anything on the inside. No front head is good, too. It's usually a drier sound, and you're usually just packing the towels against the batter head. Just put a sandbag in front to hold the towels against the head.

How about cymbals?

One thing for recording is that you probably want a heavier ride, but you don't want that heavy of a cymbal for the crashes. You also have to be careful when you mix weights.

For example, if you're using Zildjian A Custom crashes, you don't want to use a medium. You want to stay with the thins rather than try to mix in a Rock Crash with that, because the thicker cymbals are made for more of a live situation. They're made to be loud and made to cut, and sometimes they can sound a little gong-like to the mics. On the other side of the coin, if you're playing all Rock Crashes and the engineer can deal with the level, that's not so bad either, because the volume is even, but a thinner cymbal mixed in with those would probably disappear.

What records better, big drums or smaller ones?

It depends what you want your track to sound like. When I started my company, people would always say to me, "Why would someone want to rent your drums when they have their own set?" For one simple reason: Most drummers have a single set of drums. If they're going for a John Bonham drum sound, they're not going to get it with, say, a Ringo set. A lot of times when they go into the studio, the producer says, "You know, I really heard a 24-inch kick drum for this song. I hear that extra low end," but the drummer's playing a 22, so it's important to have the right size drums for the song. If you're going for that big double-headed Bonham sound, you really should have a 26. If you're going for a Jeff Porcaro punchy track, like "Rosanna," then you should probably have a 22. That's my whole approach; you bring in the right instrument for the sound you're going for. You don't try to push a square peg into a round hole.

How much does the type of music determine your approach?

The drums that I bring for a hip-hop session are actually very close to what I bring for a jazz session. Usually the hip-hop guys want a little bass drum, like an 18, and that's what's common for a jazz session, to have an 18 or a 20. Then maybe a 12- or a 14-inch rack tom, which is also similar to the jazz setup. The big difference is in the snare and hi-hats and the tuning of the kick drum and the snare.

On a jazz session, I would keep the kick drum tuned high and probably not muffled. On a hip-hop record I would tune the kick probably as low as it would go and definitely not have any muffling so it has that big "boom" as much as possible. I would also have a selection of snares from like a 4- by 12-inch snare, 3 by 13, and maybe a 3 by 14. On a jazz record, I'd probably send them a 5 by 14 and a 6-1/2 by 14. The hi-hats on a jazz record would almost definitely be 14's, where a hip-hop record you'd want a pair of 10's or 12's or maybe 13's.

Obviously it's open to interpretation, because I'm sure a lot of hip-hop records have been made with bigger sets, but when I've delivered what I just said, it usually rocks their boat.

For more on the Drum Doctors, go to drumdoctors.com.

DRUM-PREP CHECKLIST

Now that we've looked more closely at what it takes to make a drum kit sound great, here's a simple checklist to follow before you set up any mics.

- ☐ **Have the heads on all the drums been changed?** Be sure to change at least the top heads before recording.
- ☐ **Have the drums been tuned?** Are they tuned to work with the other drums in the kit?
- ☐ **Are there any sympathetic vibrations occurring?** Tune the drums so that any drum that's hit does not cause another to ring.
- ☐ **Is there an unwanted ring?** Suppress it with tape, a muffling ring, or Moongel.
- ☐ **Is the hardware quiet?** If not, spray with a lubricant like 3 in 1 oil or specialized drum hardware lubricant.
- ☐ **Is the volume level of the cymbals all the same?** Balance the level with lighter or heavier cymbals as needed.
- ☐ **Is there another snare drum available?** A song may call for a different snare drum sound.

RECORDING DRUMS

The drum kit usually gets extra focused attention in most sessions because just about all modern music is rhythm oriented and highly dependent upon the drums for the song's pulse. In fact, in most rock, pop, R&B, and country music, a wimpy-sounding drum kit equates to a wimpy track, and that's why we're dedicating an entire chapter just to recording drums.

It's a fallacy to believe that the only way to achieve a big rockin' drum sound is by miking every drum and cymbal, though. In fact, there are many tried and true methods of drum miking that have been the source of hit records for decades and that use anywhere from only one to three mics.

Whichever method you choose, try looking at the drum kit as just a single instrument, since multiple drum miking can be thought of as not much different from trying to record the individual strings on a guitar while a chord is being strummed. We're not looking so much for each discrete note, but the overall sound of the instrument.

And don't forget that the drums have to sound great by themselves first in order to sound great when recorded, and that a great drummer is a huge part of the equation.

BEFORE YOU BEGIN

As a general starting point before you even begin any complex drum miking, it's important that you get a good picture of what the drum kit sounds like. To do this, try the following:

- **Go into the room with the drums and have the drummer play the song that you're about to record.** Note the tone of the drums. *It's important that the drummer plays the song, since random playing or warming up might have a different intensity that can change the tone of the drums.*
- **Place a single mic 8 to 10 feet in front of the kit at about the same height as the drummer's head.** A large-diaphragm condenser will work nicely for this.
- **Record the kit for a minute or two.**
- **Listen to the playback.** Is the set balanced, or do one or two drums or cymbals stand out?

This will give you an idea of what the drum kit sounds like and what the issues may be when you begin recording, which now allows you to compensate by drum tuning, mic selection, or placement.

General Considerations

Here are number of things to consider that generally apply to just about any drum miking setup.

- Microphones aimed at the center of the drum will provide the most attack. For more body or ring, aim it more towards the rim.
- The best way to hear exactly what the drum sounds like when doing a mic check is to have the drummer hit the drum about once per second so there's enough time between hits to hear how long the ring is.
- Try to keep any mics underneath the drums at a 90-degree angle to the mic on top to keep the acoustic phase shift to a minimum.
- Most mics placed underneath the drums will be out of phase with the tops mics. Switch the polarity on your preamp, console, or DAW and choose the position that has the most bottom end.
- Try to keep all mics as parallel to each other as possible to keep the acoustic phase shift to a minimum.
- The main thing about mic placement on the drums is to place the mics in such a way that the drummer never has to be concerned about hitting them.
- The ambient sound of the room is a big part of the drum sound. Don't overlook using room mics where possible.

TIP: Remember that you're going for a drum sound that's appropriate for the song, not necessarily what's a good sound.

MINIMAL MIKING SETUPS

Here are a number of very simple drum-miking setups that have been used very effectively throughout the years. Remember that just because you may have a lot of mics and tracks at your disposal, that doesn't necessarily mean that you'll get a better drum sound. Sometimes, the simplest methods can provide a realism that you just can't get from close miking every drum.

Single Mic Setup

This is the simplest of all drum miking setups and can yield some surprisingly good results if you give it a chance.

Technique #1: Visualize an equilateral triangle with the base of the triangle being the overall width of the kit; then position a mic (start with a cardioid pattern but don't be afraid to try other patterns) at the apex of the triangle, directly above the center of the kit. In other words, the height of the mic is the same as the width of the kit (see Figure 8.1).

Figure 8.1: The triangle method for single-mic drum miking
© 2023 Bobby Owsinski (source: iStock Photo)

Figure 8.2: Mic in front of drum kit pointed down at a the center of the kit
© 2023 Bobby Owsinski (source: iStock Photo)

Technique #2: Position a mic 3 feet in front of the kit about 3 feet high, looking between the toms and the cymbals toward the snare. If you need more kick, lower the mic. If you need less, move it higher and away from the kick.

Technique #3: Place a mic about 3 feet in front of the drums with the mic high enough to point down at the snare (the center of the kit) at about a 45-degree angle. If you need more kick, lower the mic. If you need less, raise it (see Figure 8.2).

Technique #4: Place a mic 5 feet high and 8 feet or so directly in front of the kit. This method may provide a more balanced kit, but it will have much more room ambiance.

Technique #5: Place a mic over the drums but about even with the top of drummer's head. Make sure that the mic is angled at the whole kit in such a way as to get coverage of the toms without too much cymbals (see Figure 8.3).

Figure 8.3: A single mic over the drummer's head
© 2023 Bobby Owsinski (source: iStock Photo)

Technique #6: Place a mic over the drummer's right shoulder, angled down into the center of the kit.

- **Variation:** Use the same positioning as any of the above techniques, but use a stereo mic instead.

Figure 8.4: A typical two-mic technique
© 2023 Bobby Owsinski (source: iStock Photo)

Two Mic Setup

The two mic setup was effectively used on most records until the late '60s, so you've heard the technique on many hits from that period. While the single mic method usually lacked enough bass drum, the two mic setup provides the extra coverage needed for a more balanced drum sound. But as you'll see, there are several ways to implement that additional mic.

Technique #1: The classic two mic setup, place one mic on a short stand about 6 inches away from the front head of the bass drum, and position the second mic about a foot over the drummer's head, looking down at the middle of the kit as an overhead (see Figure 8.4). While the drummer is playing, have someone move the overhead mic around until the kit sounds balanced through the speakers. If you're not getting enough snare, for example, move it a little more toward the snare, or if you're getting too much, move it the other way. You may want to add a little equalization at 12kHz to give the kit a little more clarity and crispness.

Technique #2: Looking at the drums, place a mic on the ride cymbal side and a different model on the hi-hat side, both about 4 to 5 feet away. The dissimilar mics provide a really nice character spread from side to side, and when placed properly, provide the character of both mics in a mono playback (see Figure 8.5).

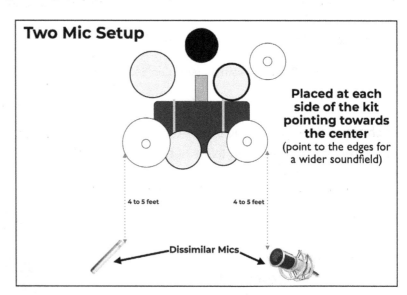

Figure 8.5: Two dissimilar mics in front of the drums
© 2023 Bobby Owsinski (source: iStock Photo)

Technique #3: Producer Brendan O'Brien's two-mic drum technique:

- Good sounding drums
- Good drummer
- AKG D 30 on kick
- Telefunken U 47 tube about 5 feet high and 3 feet in front of drums

Figure 8.6: Three-mic configuration using overhead, snare, and kick-drum mics
© 2023 Bobby Owsinski (source: iStock Photo)

Three Mic Setup

The three mic setup comes in two flavors - as an augmentation to the two mic techniques mentioned above, and as a simple stereo miking technique.

Technique #1: Add a snare drum mic to any of the two mic positions mentioned above (see Figure 8.6). Follow the snare miking techniques outlined later in the chapter.

Technique #2: Place a mic about 6 inches in front of the kick drum. Place additional mics about 3 feet high over the bell of the farthest cymbal on the left and right sides of the kit, pointing straight down at a 90-degree angle (see Figure 8.7). Move the mics as needed to get the perfect balance and stereo image.

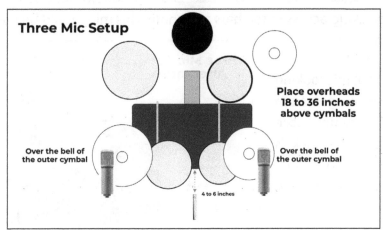

Figure 8.7: The wide stereo three-mic technique
© 2023 Bobby Owsinski (source: iStock Photo)

Technique #3: Place the first mic about 6 feet away and 18 to 24 inches off the ground in front of the kick drum. The object of this mic is to pick up not only a good bass drum sound, but some of the rest of the kit as well.

Place the second mic over the snare about a foot above the drummer's head. Place the control room monitors in mono, and listen. Move the mic until it blends with the front mic for a clear snare, kick, and open tom sound. The key here is to add that mic so you get the snare, hat, top of the toms, and cymbals without the cymbals being out of balance with the rest of the kit. You may have to move the mic closer to the drummer's left shoulder (if he's playing right-handed) to get this balance.

The third mic is placed about 6 feet from where the drummer actually hits the snare drum, but aimed so it's just peeking over the floor tom, looking at the snare. This mic will provide added depth to the kit (see Figure 8.8).

Please note: All mics should be about the same distance away from one another, which makes everything somewhat phase-coherent. If 6-foot distances are too ambient, move all the mics closer, but make sure they are all about the same distance from one another.

- **Variation 1:** Place an additional room mic 6 feet in front of the kick at the height of the top of the rim of the kick.

Technique #4: Place a mic on the beater side of the kick, about 6 inches off the floor (see Figure 8.8).

Figure 8.8: Large diaphragm mic on the beater side of the kick
© 2023 Bobby Owsinski

Place a second directional mic on the floor tom side, about two feet behind the drummer, with the drummer's body blocking access to the hats and snare. This mic should be placed just higher than the rim of the floor tom.

Place the third directional mic looking mostly at the hat and snare, again from about 2 feet behind the drummer (see Figure 8.9). Once again, the trick is to keep all three mics in a equilateral triangle, with the distance between them roughly the same.

Figure 8.9: Three mic setup behind the drums
© 2023 Bobby Owsinski (source: iStock Photo)

Technique #5: Place a dynamic mic anywhere from 1 to 4 feet in front of the kick drum. Place a pair of small-diaphragm condensers in an X/Y array about 4 to 5 feet over the dynamic mic, aimed at either the outer side of the rack toms (assuming there are two) or at the cymbals.

- **Variation:** Try one of the small-diaphragm condensers over the drummer's right shoulder. If more snare is required, add an SM57 a foot or two off the side of the drum.

Technique #6: Place a stereo mic about 5 to 6 feet over the snare drum, and a large-diaphragm condenser 5 to 6 feet in front of the bass drum. Move the stereo mic up or down to achieve the desired balance.

Four Mic Setup

The four mic setup is the beginning of the more sophisticated multi-mic setups that we use today for modern drum recording. While some of the following can be used in mono, most are designed to capture the drums in stereo.

Technique #1: Add a snare mic to any of the three-mic setups mentioned above. Follow the snare miking techniques outlined later in the chapter.

Technique #2: Close-mike the kick drum from a distance of about 1 to 2 feet in front. Close-mike the snare/hat from about 1 to 2 feet on the side looking in. Place a mic 3 feet over the rack toms and a second about 3 feet over the floor tom. The mic over the floor tom should be aimed at the floor tom from a foot or so behind the kit. This way you get a good image on the rest of the kit as well.

Technique #3: In this technique, the kick and snare mics stay the same, but the overheads are placed in a crossing ORTF configuration (see Chapter 6). Place them about a foot above the head of the drummer, directly in the middle of the kit, and move them up or down as necessary to achieve the proper balance (see Figure 8.10).

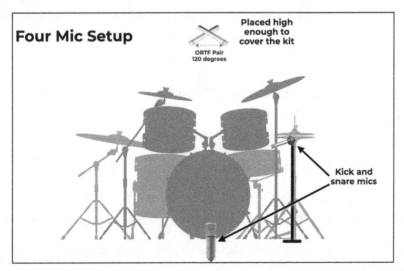

Figure 8.10: Four-mic technique with ORTF overheads
© 2023 Bobby Owsinski (source: iStock Photo)

Technique #4: Place a mic 4 to 6 inches in front of kick drum, and add a snare drum mic using the method of your choice. Place a mic 18 to 36 inches above each outer cymbal aiming at its bell (see Figure 8.11).

Figure 8.11: Four-mic technique using cymbal mics

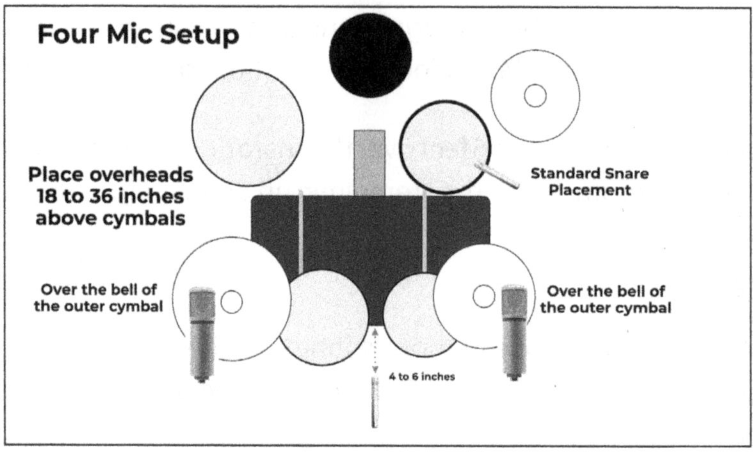

© 2023 Bobby Owsinski (source: iStock Photo)

MIKING THE DRUMS INDIVIDUALLY

Although the minimal mic setups work perfectly well, today we look for the sound and additional control that individual drum miking can provide. Here are multiple techniques to mike every drum in the kit. Keep in mind that even if a mic brand or model is mentioned, it's perfectly acceptable to use another model, so feel free to experiment.

Miking The Kick

The kick (bass) drum provides the pulse of the song, and through the years has gained more and more importance in the final mix, especially in electronic genres like EDM. Here are a number of different miking techniques, as well as some general considerations before you choose your technique.

Considerations

- A really large kick (such as a 26-inch) can have fundamentals lower than what the room can support, so you end up hearing the octave above the fundamental instead of the fundamental itself. This can put the drum's perceived pitch even higher than what you'd hear with a smaller drum.
- Shredded newspaper can work very well as damping material.
- A folded pillowcase or bath towels tucked inside the drum in front of the beater can work well for damping. One or two inch acoustic foam lining the shell can also be used for dampening.
- This is very old school, but a felt strip across the head can be pulled tight or loose to vary the amount of head dampening.
- If the kick rings too much, try putting a small pillow inside and vary the damping by varying the amount of pressure it has against the beater head.

- A good way to cut a hole into the front head of the kick drum is to heat up a saucepan lid until it's red-hot and then drop it onto the head. This burns through cleanly and doesn't leave any rough edges that can split.

- When using condenser microphones to close-mike a kick, be aware that you risk overloading the mic's diaphragm or its internal preamp. This can fool you because it can sound similar to a blown speaker.

- For more attack on any drum, add a few dB at 8kHz. For a larger-sounding kick, add a little between 50 and 100Hz. Attenuate at 1.5kHz to decrease the "honkiness." Attenuate between 200Hz and 500Hz to eliminate the "beach ball" sound.

- A limiter can smooth out the peaks, but use it sparingly when tracking. A dB or two with a 2:1 ratio, a very fast release, and 3 to 7ms attack time is many times all that's needed.

Tip: When using a condenser mic on any position of the drums, always engage the attenuation pad to decrease the chance of an overload, if the mic has one.

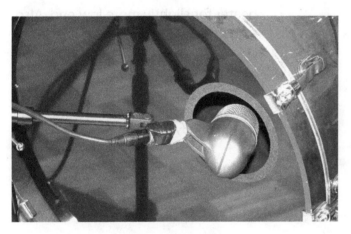

Placement

Technique #1: If the kick has a hole, place the mic just at the edge, angled at 30 to 45 degrees off-axis aimed at the beater (see Figure 8.12)

Figure 8.12: Shure Beta 52 just inside hole angled at a 45-degree angle
© 2023 Bobby Owsinski

Technique #2: If the kick has no hole in the front head, place the mic about 3 to 4 inches in front of the drum at about the same level as the beater. Move it off-center to change the sound (see Figure 8.13).

Figure 8.13: Miking a kick with a front head
© 2023 Bobby Owsinski

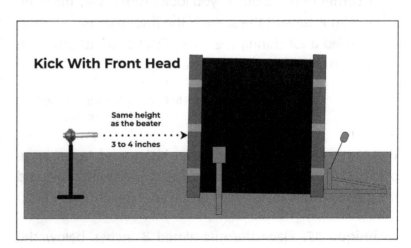

Technique #3: Put your hand in front of the bass drum while the drummer hits quarter notes. You will feel a shockwave projecting from the head of the drum. Continue to move your hand until the shockwave almost disappears. Place the mic at the edge of the shockwave, aimed at the center of the drum (see Figure 8.14).

- **Variation:** This technique can also be used to add a second kick mic (sometimes called an "out mic," as opposed to the "in mic" inside the drum) to capture the fundamental of the kick drum that a mic positioned closer to the beater might not capture. This mic also might require a tunnel (see the upcoming "The Kick Tunnel" section) in order to isolate the sound of the kick from the rest of the kit and any other instruments playing in the room.

Figure 8.14: Finding the kick-drum shockwave
© 2023 Bobby Owsinski

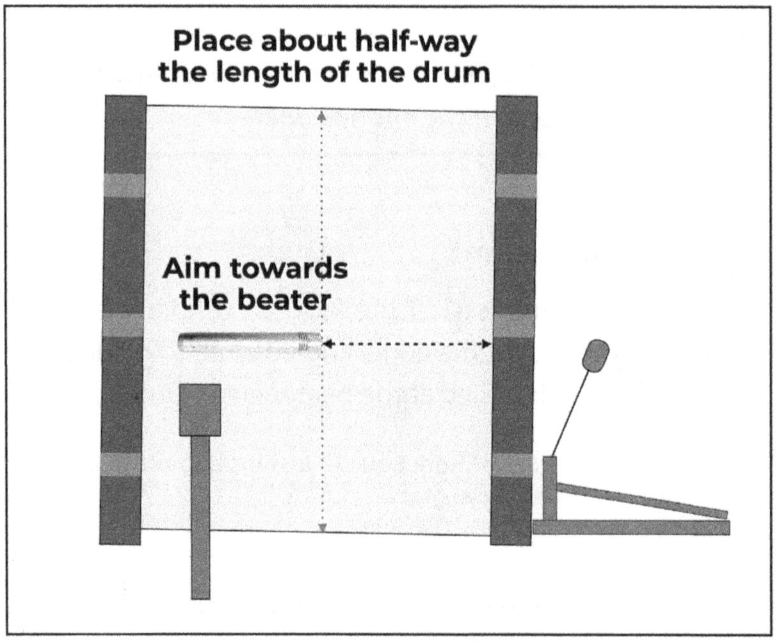

Technique #4: If the kick has no front head, place the mic about halfway inside the drum shell, pointed at the beater. Move it closer to the beater for more punch and farther from the beater for slightly more low end. Also, move the mic off-center for a different tone or to adjust the amount of punch from the beater (see Figure 8.15).

Figure 8.15: Mic just inside the kick drum
© 2023 Bobby Owsinski

Technique #5: For more tone from the drum shell, angle the mic at a 30 to 45 degree angle aimed more toward a left corner of the drum as you look inside, away from the snare. Aim it toward the snare if the floor tom is very loud and is used a lot during the song. Start at a distance of 6 inches and then adjust to taste (see Figure 8.16).

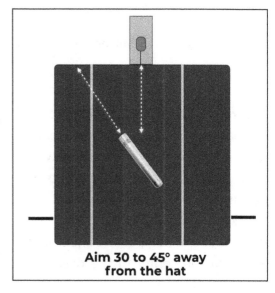

Figure 8.16: Mic angled inside the kick
© 2023 Bobby Owsinski

Technique #6: For a very aggressive kick sound, tape a large coin to the drum head so that the beater strikes the coin and not the head. (Make sure you consult with the drummer first.)

Technique #7: Place the mic aimed 3 inches below the beater. This captures some of the attack without getting too "clicky." For more click, aim it more towards the beater.

The Recording Engineer's Handbook - 5th edition

Technique #8: Place a ribbon mic anywhere from 1 to 6 feet from the outside of the kick drum. Be sure to point the diaphragm at a 45-degree angle, aimed at the floor (see Figure 8.17).

Important: If you don't use the 45-degree angle, you'll likely damage the ribbon due to the air blast from the kick.

Figure 8.17: Royer R-121 at a 45-degree angle outside the kick
Courtesy of Royer Labs

Technique #9: If a mic has been placed inside the drum, add a second one 2 to 3 feet away from the edge of the kick-drum shell, pointing at the beater. Move the mic closer or farther back to adjust the amount of room and low end it captures.

Technique #10: For a punchier sound with a doubled-head kick drum, take the front head off and then build a cradle for the mic using thin wire or fishing line attached to the mounting hardware inside the drum. Place the mic on the cradle and adjust it so it's at the height of the beater. De-solder the female XLR end of a mic cable and thread the bare end of the cable inside the drum through the air hole. Re-solder the XLR to the cable and attach to the mic (see Figure 8.18). Reattach the front head of the drum. The drummer will now feel the back pressure from having the front head, and the sound will be similar to that of a kick without the front head.

Figure 8.18: A makeshift cradle inside the kick drum
© 2023 Bobby Owsinski

Technique #11: For additional attack, add a second mic to the beater side of the kick. A small lavaliere or a clip-on condenser works well. Balance this sound against the sound from the front or inside mic.

The Kick Tunnel

To get more isolation for an outside kick microphone, a "tunnel" is sometimes constructed around it. This can be made of anything from packing blankets draped over chairs and mic stands (#1 below) to something more formal (#2 below).

Considerations

- If you have a small room where the reflected sound into the kick mic doesn't sound that good, the tunnel can eliminate those room reflections.
- One side effect of the tunnel is a slight lowering of the resonant frequency of the drum, since the tunnel will acoustically couple with it. The drum may have to be retuned up just a little to

compensate. The tone is widely variable simply by adjusting the distance between the drum and the opening of the tunnel.

Placement

Technique #1: You can construct a makeshift tunnel using chairs, mic stands, and packing blankets (see Figure 8.19). The upside of this is that the outside leakage will be reduced somewhat into the outside kick mic. The downside is that you won't be able to take advantage of the resonant qualities of a more permanent tunnel, as described below.

Figure 8.19: A kick tunnel made of blankets
© 2023 Bobby Owsinski

Technique #2: From an industrial paper-tube manufacturer (or with cardboard concrete form tubing), obtain tube sections that are 24 inches in diameter, and if possible, only 3/4 round so they're flat on the bottom. Line the tubes with 2-inch acoustic foam or Rockwool, and cap off the end with an 1/8-inch circle of Plexiglas, which is also covered inside with acoustic foam. The foam makes the tube extremely dead inside and also lowers the resonant frequency. The tube should be about 6 feet long. Use a packing blanket to close the gap between the kick drum and the tunnel.

The Subkick

The subkick phenomenon started due to the desire to get more bottom end from the kick without having to increase the low-end EQ. The unit is only meant to capture from 20Hz to 100Hz or so. The idea became popular when engineers would make homemade units utilizing the woofer from a Yamaha NS-10M and use the magnet of the speaker to attach it to a mic stand or drum hardware (see Figure 8.20).

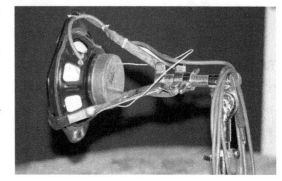

This method isn't used as much with the advent of subharmonic synthesizer plugins like Plugin Alliance Boom and Waves Submarine and MaxxBass.

Figure 8.20: A homemade subkick
© 2023 Bobby Owsinski

Considerations

- To use the subkick, get the sound of your main kick drum first and then add about 10 percent of the subkick to the main kick, or just about when you begin to hear it. Of course, if your speakers won't reproduce anything as low as 30 or 40Hz, then you probably shouldn't even try this because you'll just be guessing about how loud the subkick should actually be, and if you add too much the kick will sound too woofy and will lack definition.

- You can use the subkick on other instruments as well. Try it on a bass amp or even a trombone.

Placement

Technique #1: Place it about 2 inches from the outer head of the kick drum, or if the drum only has one head, from the edge of the shell (see Figure 8.21).

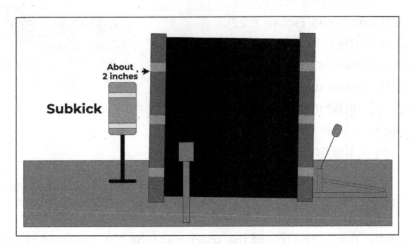

Figure 8.21: Subkick placement
© 2023 Bobby Owsinski

Miking The Snare

Regardless of which technique or mic you choose for miking the snare, the ultimate sound still comes down to the sound of the drum and the drummer. That said, the correct placement can augment the sound quite a bit, making it fit better in the track while isolating it from the other drums.

Considerations

- The Shure SM57 has been the standard snare microphone for years and shows no signs of being replaced. It's one of the few mics that maintains its punch and sound regardless of how hard the drummer is playing, thanks to its ability to take high sound-pressure levels before distorting. That said, don't be afraid to try other microphones. Many engineers get great results from small-diaphragm condensers, although they don't take the sound pressure level that a dynamic microphone will.

- The "crack" of the snare doesn't necessarily come from the top mic. For more crack from the snare, try a well-placed room mic or an under-snare mic. If there's too much cymbal or kick leaking into it, key it from the snare track during the mix.

- Add a dB or two at 10k to 12kHz for a more crisp sound.

- To smooth out the peaks when tracking, consider using a slight bit of compression (just a dB or so) with a 2:1 ratio set to very fast attack and release times.

Placement

Technique #1: This has become the "standard" snare setup when using an SM57. Place an SM57 on a boom stand and position it about 1 inch, or about two of your fingers, above the rim. The silver "ring" of where the mic head meets the body should be placed just over the rim of the drum. Make sure it's at a slight angle so the mic is pointed toward the center of the drum head (see Figure 8.22). To get some isolation from the other drums and cymbals, try to place the snare mic so its pointing away from the hi-hats, but make sure it's not in a place where the drummer will hit it. A good place is directly between the rack tom and hi-hats as long as it's out of the way of the drummer.

Figure 8.22: The standard SM57 snare setup
© 2023 Bobby Owsinski

- **Variation:** Place the mic at the rim of the snare near the hat, elevated about 2 inches over the snare. Place the mic so that it looks across the head aiming for the far edge. Adjust outward for more shell or inward for more impact (see Figure 8.23).

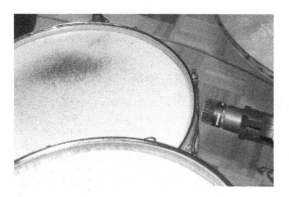

Figure 8.23: An SM57 across the snare top
© 2023 Bobby Owsinski

Technique #2: Place the mic so it's about 6 inches above the rim of the snare drum. Aim the mic at a 45-degree angle at the center of the snare drum. This technique may pick up more of the hi-hat, but it can sound more natural than Technique #1 and can be a better choice if the drummer is using a sidestick.

Technique #3: Position a mic 4 to 8 inches away from the snare and aim it at the shell. Move it closer to the bottom head for more snare sound, and closer to the top head for more attack and less buzz (see Figure 8.24).

- **Variation:** Aim the mic at the port on the side of the drum. Miking the port will give you a good, solid transient with both heads in phase. Be sure to mike the port at an angle, or you might pick up a wind blast.

Figure 8.24: SM57 placed on the snare's side
© 2023 Bobby Owsinski

Technique #4: Along with a top mic, place a second mic about 3 inches or so from the bottom head and right under the snares. Try to place the mic at a 90-degree angle from the top mic, and remember to flip the phase on the bottom mic and choose the position that has the most low end when blended with the top. Mix the bottom mic in for presence. Cut 50Hz to 100Hz from the bottom mic or use the roll-off on the mic to decrease the amount of leakage from the kick drum (see Figure 8.25).

Figure 8.25: Sennheiser 441 under the snare
© 2023 Bobby Owsinski

Technique #5: Add a second mic 18 to 24 inches away, looking in from the side. Use a hyper-cardioid pattern if available.

Technique #6: For additional snare presence, place a condenser mic in the room at a point where the snare sounds great, and print it to another track. Be sure that the pad and high-pass filter are both switched on. During the mix, put a gate on that track with a key function triggered by the original snare signal, so that it only opens up when the snare hits, and adjust the parameters to taste.

Technique #7: For better isolation from the hat, cut off the top of a plastic milk jug or bottle down to the end of the handle. Slip the snare mic backwards into the hole and then into the mic clip (an SM57 fits just right). Be aware that this will most likely change the sound of the microphone (probably for the worse), but you'll increase the isolation (see Figure 8.26).

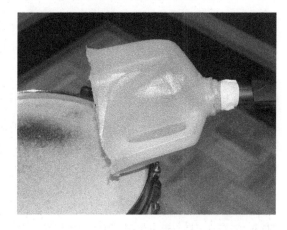

Figure 8.26: Plastic bottle used to isolate the snare mic
© 2023 Bobby Owsinski

- **Variation:** Use a Styrofoam or paper drinking cup instead of the plastic jug or bottle (see Figure 8.27).

Figure 8.27: Drinking cup used to isolate the snare mic
© 2023 Bobby Owsinski

Technique #8: Tape a small-diaphragm condenser to an SM57, taking care to exactly line up both capsules to ensure that the mics are in phase. The condenser will provide the top end, and the 57 will provide the body to the sound.

TIP: There are now dedicated dual microphone adapters available from Shure and Wilkinson Audio that will make using two mics on the snare much easier.

Technique #9: Take a contact mic, such as a Barcus Berry or Korg, and tape it to the top of the snare drum head, but out of the way of the drummer. Connect it to a mic pre and then to the key input of a gate. If you use a gate with a filter section, you will be able to remove all other frequencies from the key input. The gate will then open on every snare hit and should supply additional attack to the sound.

- **Variation:** Place the contact mic on the bottom head of the snare or even on the drum shell.

Brushes

Brushes aren't used that often, but when they're needed it's good to know a few techniques to adequately capture the sound.

Considerations

- The brush sound is partly the attack of the individual bristles (which you want to capture so it sounds crisp), partly the ring of the rim and shell, and partly the ambient sound from around the kit.
- Getting a good brush sound comes from how you mike the entire kit, not just the snare. Close-miking the snare may not work for this application, so it's better to think in terms of the sound of the entire kit.
- Part of the trick is to use mainly the overheads, as that provides the depth to the sound.
- Coated heads are recommended for the best sound.

Placement

Technique #1: If you're using overheads over the drums instead of cymbal mics, move them in closer to about head height of the drummer. Adjust up or down as needed.

Technique #2: Place a dynamic mic about an inch from the shell of the snare, pointed at the center of the shell. Move up or down to adjust the attack and mix with the overheads for realism.

Technique #3: Tape a piece of cardboard to the top of the snare and have the drummer play the brushes on that. Sometimes it can sound more like a snare than a real snare.

- **Variation:** Play with brushes only on a cardboard box.

Miking The Hi-Hat

Sometimes the snare mic picks up so much of the hat that many engineers fail to see the point of using a dedicated hi-hat mic. Even with a large amount of hat leakage already bleeding into the snare mic, giving the hat its own mic and track is worth it just for the additional control it provides, such as the ability to increase the level in certain sections, bring out ghost notes, or to enhance the sound through equalization or effects.

Considerations

- Heavy hi-hat cymbals that sound great for live work tend to have a lot of low overtones that cause frequency interaction with some of the drums when it comes to recording. Lighter hats tend to record better as a result.
- Placing a mic too close to the end of the cymbal might pick up the air noise, while placing it too close to the bell might capture a sound that's too thick.
- A condenser mic is usually used to mike the hi-hat to best capture its transients.
- Be sure to use a -10dB pad on the hat mic if it's available, since hi-hats (especially heavier ones) can sometimes be quite loud.
- Make sure that the hat mic is placed on the far end of the hat, as far away from the crash cymbal as possible for maximum rejection.
- Add a dB or two at 10 to 12kHz to give the hat sizzle.
- Sometimes it's best to filter out everything below 160kHz or so using the high-pass filter on the console or preamp. These frequencies won't be missed and will tend to clean up the sound when the rest of the drums and cymbals are placed in the mix.
- If the hi-hats sound too thick or heavy, attenuate 1.2kHz by 2 or 3dB, or move the mic away or more toward the edge of the top cymbal.

Placement

Technique #1: Place a mic pointing straight down at the cymbal, about halfway from the edge of the rim, but placed toward the back of the hat. Keep it away from where the drummer is hitting for more isolation from the rest of the kit (see Figure 8.28).

TIP: Move the mic closer to the bell for a thicker sound and more toward the edge for a thinner sound.

Figure 8.28: Mic on the outside of the hi-hat
© 2023 Bobby Owsinski

- **Variation:** For maximum isolation from the rest of the kit, place the mic midway from the back end of the hat, but angle it at a 45-degree angle away from the kit.

Technique #2: Position the mic about 4 to 6 inches above the hat and angle it toward the place where the drummer hits the hat. This is where you get the most clarity and attack, although you'll also pick up more leakage from the kit. If you need more air and sizzle, move the mic higher up and aim it straight down toward the cymbal (see Figure 8.29).

Figure 8.29: AKG 452 on the inside of the hat
© 2023 Bobby Owsinski

Technique #3: Place a mic looking down the post of the hi-hat for an extra thick sound. Many times a lavaliere mic can work great in this application, since it will provide plenty of isolation from the rest of the kit.

Miking The Toms

When musicians, producers, and even some drummers refer to the sound of a kit, many times what they're referring to is the sound of the toms. Even if the rest of the kit sounds great, a kit can still get low marks for wimpy-sounding toms. As with the rest of the kit, make sure that the toms sound great acoustically before the mics are placed in order to get the best sound.

Considerations

- Although dynamic mics are often used on toms, condenser mics capture the transients of the instrument better. As a result, you'll hear more of the attack of the drum and less of a thump using a condenser mic rather than a dynamic.

- Sometimes placing the mic between the tuning lugs can get a truer sound from the drum than if it's placed over a lug.

- Placing the mic too close to the drum head will increase the attack that you capture, but at the expense of capturing its body, and therefore the tone, of the drum.

- When miking multiple toms, try to keep all the tom mics facing in the same direction as much as possible to eliminate any possible phase issues between them. If you were looking down at the drummer from the ceiling behind the drum kit, they would all be pointing to the six o'clock position towards the drummer.

- With most toms, adding a dB or two of EQ at 5kHz can emphasize the attack and at 8kHz can emphasize the presence. Attenuating 1.5kHz can decrease the thickness. Attenuating 400 to 500Hz may decrease any "boxiness" that the drum might have, but too much attenuation can also suck the tone out of the drum sound.

- Top and bottom tom miking can sometimes be a good approach, but the technique can be very phase-sensitive (check the phase again) and generally lacks the clarity of a single top mic.

- When miking a very large kit with a lot of toms, sometimes miking each pair of toms may work better than miking each one individually.

- Use a small amount (1 inch by 3 inches) of console or gaffer's tape to dampen a resonant ring in the drum sound without ruining too much of the snap of the drum. Tape it about 3/4 of an inch from the edge of the top head, pull it back a bit, and then stick it down over the hoop on the outside of the drum. Done correctly, it should stick onto the top head for only about half of the tape's length. "Deadringers," which are 1/2-inch-wide rings of thin plastic that go over the periphery of the top head, do a similar job but can eat up too much of the drum's tone. Another substitution to try is Moongel.

- The under-tom approach where the mic is actually placed inside the drum may be helpful on top head-only toms (concert toms) where leakage is a problem.

- The two heads of a drum can usually be tuned in such a way to give the drum much more tone and power than a single head alone.

- Gates can be used to diminish the ring from a tom that you can't tame with tape, padding, or tuning, but use them only during mixing, not during recording. Remember, the leakage into your tom mics lends to the overall sound of the kit. If you gate them, you might destroy your snare sound, too.

Placement

Technique #1: The classic method: Place the mic 2 to 3 inches off the head above the rim (about three fingers on edge) at a 45-degree angle, looking down at the center of the head to get the most attack (see Figure 8.30). For more ring and less attack, point the mic closer to the rim.

Figure 8.30: U 87s over each rack tom aimed away from each other for phase aligment
© 2023 Bobby Owsinski

Technique #2: Instead of miking a tom from above, try placing the mic a few inches underneath the drum.

Technique #3: Place a top mic about 4 inches over the top head and a bottom mic about the same distance away from the bottom head. While listening on the monitors to the drummer playing slow quarter notes, have the bottom mic moved in closer until the sound aligns with the top mic. This is important because the attack reaches the bottom mic just a bit later, so moving it closer will more precisely align the attack reaching both mics at the same time. This technique also has the advantage of canceling out some of the room reflections. Be sure to check the phase.

Technique #4: On a floor tom, place the mic anywhere from 2 to 4 inches over the drum head and pointing at about a 45-degree angle aimed at the center of the head (see Figure 8.31).

- **Variation:** If there's not enough bottom end even after tuning, add a mic with a strong low end (like an AKG D112) under the drum. This mic will pick up only the low-frequency information, but it may also have a noticeable time delay compared to the top mic. Flip the phase switch on your preamp, DAW, or console and select the position with most low end.

Figure 8.31: MD 421 on a floor tom
© 2023 Bobby Owsinski

Technique #4: You can sometimes get the greatest rejection of the rest of the kit from the floor-tom mic by placing it underneath the ride cymbal, again about 3 inches above the rim, pointed at the center of the head (see Figure 8.32).

Figure 8.32: Floor miked from underneath the ride cymbal
© 2023 Bobby Owsinski

The Overhead Mics

Depending upon the sound you're going for (which is dependent upon the song, artist, arrangement and player) and the environment you're recording in, the overhead mics either can be used to capture the sound of the entire kit or can be used primarily as cymbal mics. That's why there's such a variety of miking positions and philosophies when it comes to overheads. Regardless of the technique you choose, pay close attention to the sound that the overheads supply, because it will have a great bearing on the overall sound of the drum kit.

Considerations

- Make sure that the -10dB pad is inserted either on the mic or on the console/DAW interface, since the overhead mics normally generate a high output.

- Generally speaking, the image is clearer with an X/Y overhead configuration, and there are fewer phase issues when using the overheads that are intended to capture the sound of the entire kit.

- If you have a room that's too live, move the overheads closer to the kit to reduce the amount of room being picked up.

- Sometimes you can clarify the sound of the cymbals and clean up the ambient sound by engaging the hi-pass filter either on the mic or on the console.

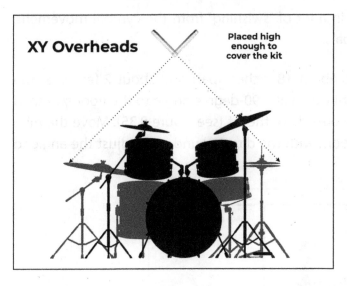

Figure 8.33: Overhead miking using the X/Y configuration
© 2023 Bobby Owsinski (source: iStock Photo)

- Make sure the kit sounds balanced when listening to only the overheads if they're used to capture the entire kit. The snare should be just right of center, the same as if you're looking at the kit (if you're mixing the kit from the audience perspective).

- If you're using the overheads only as cymbal mics, check the cymbal balances. Move the mic away from a cymbal if it's too loud and closer to it if it's too quiet.

Placement

Technique #1: To use the overheads to record the overall drum kit, position an X/Y microphone pair about 2 feet over the drummer's head, centered over the middle of the kit (see Figure 8.33). Raise or lower the mics to achieve the desired kit balance, as well as the balance between the amount of room sound and the direct sound of the kit.

TIP: Lower the overhead pair if you're recording in a room with a low ceiling, since the reflections off the ceiling will probably have an undesirable effect on the sound.

- **Variation #1:** Instead of the mics configured as an X/Y pair, position them in an ORTF configuration.

- **Variation #2:** Use a stereo mic instead of a stereo pair for easier positioning.

- **Variation #3:** Position the stereo mic or microphone pair about 2 feet behind the drummer, aiming at the top of the cymbals.

Technique #2: For cymbal miking, position a mic from 2 to 3 feet above the outside cymbal on each side of the kit (see Figure 8.34). Move the mics lower for more cymbals and less room, and higher for more room. Also move the mics as needed to be sure that all the cymbals are heard at the same level.

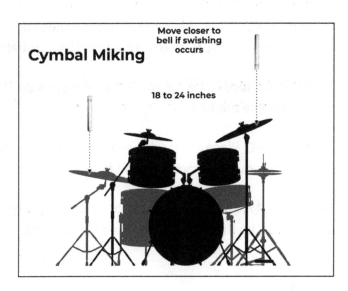

Figure 8.34: Cymbal miking
© 2023 Bobby Owsinski (source: iStock Photo)

137 | Recording Drums

- **Variation**: If the above placement results in a lot of "swishing" from the cymbal movement, place the mic closer to the bell of the cymbals.

Technique #3: Place the mics over the rack toms about 18 inches apart and about 2 feet over the drummer's head. Start with the mics aimed straight down at a 90-degree angle to the floor and then point them at a 45-degree angle back toward the outside of the kit (see Figure 8.35). Move the mics higher or lower to balance the ambiance of the room with the direct sound, and adjust the angle to balance the cymbal level.

Figure 8.35: Centered overhead mics at a 45-degree angle
© 2023 Bobby Owsinski (source: iStock Photo)

Technique #4: Place an X/Y configuration as high as you can (stay at least 12 inches away from the ceiling), aimed toward the very outside of the cymbals on each side of the kit. Change the angle so they're aimed more at the bell of the cymbals for a fuller tom tone and a narrower stereo separation.

Technique #5: If the ride cymbal is too quiet compared to the other cymbals, place a separate mic about 6 inches away from the middle of the cymbal. Adjust the level to taste.

- **Variation**: Place the mic about 6 inches underneath the ride, pointed at the bell. Make sure to reverse the phase!

Technique #6: Place a mic with an omni pattern above the cymbals on the hi-hat side of the kit and a cardioid mic over the cymbals and floor tom, angled slightly out toward the room.

Technique #7: For really old-school overhead placement, place a single ribbon mic a foot over and to the right of the drummer's head (see Figure 8.36).

Figure 8.36: Single overhead mic
© 2023 Bobby Owsinski (source: iStock Photo)

Four Ways To Tame Cymbals That Are Too Loud

We've all been there. You're either on stage or in the studio with a drummer that has no idea that his or her cymbals are way louder than the rest of the kit. Everything's fine until you come to a tom fill or chorus when either the crash (the usual culprit) or the ride cymbal suddenly wants to drown everything out in the room. Here are four ways to attempt to get those cymbals under control.

1. **Change the cymbals.** Let's start at the source, which is the cymbals themselves. As stated before, many drummers choose cymbals for the wrong reason and don't try to match them to the situation. Heavy cymbals will cut through a loud band on stage and are hard to break, but put them in the studio and they'll wash out a drum kit in no time. Generally speaking, lighter cymbals sound better in the studio and are a closer match to the volume level of the drums.

2. **Ask the drummer to hit less aggressively.** A great drummer has a feel for how aggressive to be with cymbal hits so the volume level is appropriate for the section of the song. In that case, if the level is too loud you can easily say, "Can you back off on the cymbals a little?" and know that you've accomplished the level change without have to do much else. Some drummers just can't do this however, either because they're not capable of that kind of internal command or feel that it somehow decreases the passion in their playing.

3. **Control the level of the cymbals**. There are a lot of ways to control the level of cymbals themselves, ranging from old-fashioned gaffers tape to products like Drumtacs, but level control usually comes at a cost of tone. It's a cheap and fast way to accomplish the task, but more appropriately weighted cymbals are the better choice.

4. **Lower the gain on the cymbal or overhead mics**. This is a last resort since it really won't work all that well. The problem with cymbals that are too loud is that they'll bleed into the other drum mics (especially the toms) so you're still left with the same problem, only it might sound even worse.

Best case scenario is that you have a drummer that's in tune with his or her cymbals and aware that their volume level is important to the overall sound of the band. Otherwise, the other fixes are just band aids that probably won't give you the results that you're looking for.

The Room Mics

Sometimes room mics can be the added glue that makes the individual drums sound like a single coherent drum kit. As you'll see in some of the following techniques, the room mics don't have to be far away from the drum kit to achieve the desired result.

Considerations

- Consider what you're trying to achieve when recording the room. If the goal is more ambiance, then a single mic pointing away from the drums might work well. If the goal is to get a bigger drum sound, then place a mic or mics at the point in the room where the kit seems most balanced.

- A figure-8 pattern might work better used as a room mic in a small room because the mic will pick up the reflections from the back wall, but not the side-to-side or floor-to-ceiling reflections that sometimes sound bad. Even if the mic is angled it will still pick up fewer of these reflections than any other pattern. As a result, whereas a small room may add an unpleasant sound when recorded with omnis or cardioids, it may sound perfectly acceptable with a figure-8.

- Generally speaking, the fewer mics used to close-mike the drums, the more effective the room mics will be.

- Two or three dB at 12k to 14kHz can open up the room sound and give it a little "air." Likewise, a high-pass filter set to between 100 and 150Hz can attenuate the low frequencies that make the kit sound muddy or boomy.

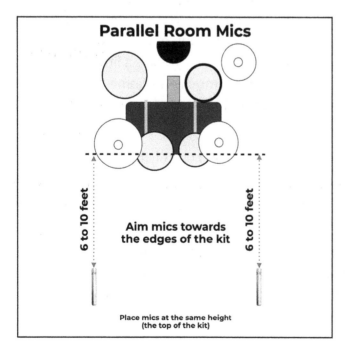

- If there's a lot of cymbal activity in the song, place the room mics closer to the floor for a warmer sound.

Placement

Technique #1: Use two mics, each placed 6 to 10 feet away from the kit and pointed at the outer edge of the farthest cymbal at each side. The mics should be at the same height and be exactly parallel to each other (see Figure 8.37).

Figure 8.37: Parallel room mics
© 2023 Bobby Owsinski (source: iStock Photo)

The Recording Engineer's Handbook - 5th edition | 140

Technique #2: Place a stereo mic 6 to 8 feet in front of the kit at about 7 feet high. The mic should be at about a 45-degree angle, facing down at the kit.

Technique #3: Have the drummer hit only the snare and find a point in the room where it takes on the character of the "crack." Place a mic at that point. When mixing, gate the room mic track using the close-miked snare to trigger the gate via a sidechain. Remember that adding compression (sometimes in large amounts) changes the character of the sound.

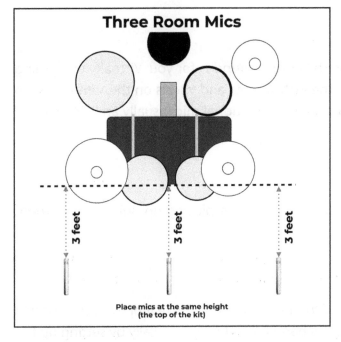

Technique #4: Place two mics about 3 feet away from the farthest outside edge of the left and right cymbals. Add a third mic placed in the center of the kit. Set them at a height of the drummer's head (see Figure 8.38). Point the center mic down at the snare but adjust it to taste. Pan the outside mics hard left and right and the center mic to the center.

Figure 8.38: Using three room mics
© 2017 Bobby Owsinski (source: iStock Photo)

- **Variation #1:** Just use the two outside mics.
- **Variation #2:** Just use the center mic.
- **Variation #3:** Place the room mics down low, 2 feet off the floor and pointed upwards (see Figure 8.39).

Figure 8.39: U 47 as a drum-kit room mic
© 2023 Bobby Owsinski

Technique #5: Place a midfield mic from 3 to 6 feet away from the kit, but placed halfway between the kick and hi-hat. Move the mic backward and forward to get the best balance. Move the mic toward the kick if it's not being picked up enough.

- **Variation #1:** Add two additional room mics as in Technique #1.
- **Variation #2:** Switch the main mic to omni.

Technique #6: Turn the room mics away from the drums and point them at the room to pick up more room ambiance and slap from the walls.

DISTINCTIVE DRUM SOUNDS

Sometimes we can pick out either the artist or an era in which a song was recorded just by the sound of the drum kit, since drum recording has evolved so much from the '50s until now. Here are a number of tricks to take your sound back to those times if it works for the song you're working on.

The '50s Drum Sound

If you really want to get adventurous, then the '50s style drum sound is for you. Virtually everything recorded in during the 1950s was in mono with all the instruments and vocals on the same track. In fact, some of the biggest hits of the era used just a single bi-directional mic (usually an RCA 44 or 77 ribbon mic) to pick everything up.

Considerations
- The sound of the room is critical, since the single mic will be picking up quite a bit of room reflections along with source signals.

Placement

To record this way, the front of the figure-8 mic is used for the vocal, while the rear of the mic is used to pick up the instruments in the band. The balance of the band is created acoustically by setting up the instruments either closer or further away from the mic in order to achieve the proper balance.

The '60s Beatles Sound

The Beatles' drum sound changed quite a bit through their eight years of recording. Ringo's kit used mostly minimal miking in the beginning, as was the standard of the day, then gradually evolved to a setup somewhat similar to what we normally use today on the very last song the band recorded (appropriately titled "The End").

Considerations
- The drum sound on later Beatles records and on songs such as "Hey, Jude" and "Come Together" utilized tea towels (thin towels similar to cloth napkins) covering the heads of the snare and toms. The front head of the kick drum was also removed, and the sound was dampened with towels, a radical move for the time but commonplace today. The sound is enhanced by the use of a Fairchild 660 limiter.
- Regardless of the number of microphones used on the drum kit, the drums were always recorded in mono to only a single track, and in the beginning, even sharing the track with other band instruments. The only exception was on the aforementioned "The End," which was recorded in stereo on two tracks.

Placement

Technique #1: For the early Beatles drum sound, place a cardioid mic (an AKG D19 was originally used) over the center of the kit at a height of around the top of the drummer's head. Move it closer to the snare if more snare sound is needed. Place a large-diaphragm cardioid mic (an AKG D20 was originally used) about 3 to 6 inches from the kick drum, raised to about the level of the top of the rim and pointed downward at a 45-degree angle.

Technique #2: For the later Beatles drum sound, augment the above with a cardioid mic under the snare from about 6 inches away from the edge of the drum and pointed up at a 45-degree angle. Do the same for the rack tom. For the floor tom, place an identical mic 6 to 9 inches away from the top head. Don't forget the towels over the heads.

- **Variation:** During the *White Album* sessions, engineer Ken Scott moved the mics from underneath the drums to the top as we use them today. The tea towels remained in place, however.

The '70s Drum Sound

The late '60s/early '70s drum sound was a more dead and a tighter sound that what had previously been the norm. In fact, even studio design followed suit, with the trend of the day being very dead with no reflections. Thankfully, at least that fashion has since died in most parts of the world.

Considerations

- Although a dead sound is what you're going for, a common mistake is to deaden the drums too much.
- The reason for the damping was to kill any ring so you would just hear the attack of the drum.

Placement

Technique #1: Deaden the snare drum by placing a thin towel over it. Place the two overhead mics about 18 to 24 inches directly over the cymbal bells. Deaden the toms and kick enough to eliminate most of the ring so you hear only the initial attack when the drum is struck.

- **Variation #1:** Remove the batter head and rim of the drum and place a strip of felt or cloth 2 to 3 inches wide completely across the drum. This is the classic way that drummers were taught to dampen drums in the past.
- **Variation #2:** Cut the rim off an old snare head and then put the remaining part on the drum head to deaden it.
- **Variation #3:** Cut off both the rim and the inside of an old head and place it on the drum. Make sure that you use a little tape to hold it in place, or the stick will get caught up in it when the drum is struck.

- **Variation #4:** Tape half of a sanitary napkin to the drum head on one side only so the other side flaps up on the attack, then settles back on top of the drum head. A pad can also be applied to the bottom head if the ring is too long.

- **Variation #5:** Tape a piece of cloth to one edge of the snare with a flap about 3 or 4 inches over the top head. Place a man's wallet under the flap (see Figure 8.40).

Figure 8.40: Wallet on the snare-drum head
© 2023 Bobby Owsinski

Technique #2: For that old Motown disco hi-hat sound, mike the top hat cymbal by pointing a mic directly at the bell from about an inch away (see Figure 8.41). You can also use a small lavaliere mic taped directly to the top of the stand.

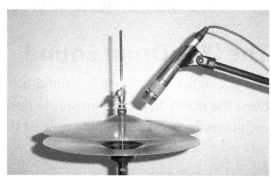

Figure 8.41: A mic aimed at the hi-hat bell
© 2023 Bobby Owsinski

The Led Zeppelin Sound

Drummers and producers have admired the big roomy sounds of Led Zeppelin drummer John Bonhom for decades now, and the sounds are easier to get than you think since the miking setup is minimal. What's especially interesting is the fact that Bonham's kit was the first example of stereo drums on record, and it was an accident!

According to engineer Glyn Johns, the sessions for the first Led Zeppelin album began with a simple two microphone drum setup - a single overhead mic placed 4 to 5 feet over the snare drum, and a second mic placed in front of the outer head of the kick drum. While recording, Johns asked his assistant to make a change on guitarist Jimmy Page's amplifier, who moved the microphone to access the controls. He forgot to move it back in front of the amp's speaker cabinet however, and it wound up pointing across the drum kit on the floor tom side. When Bonham began to play everyone in the control room was astounded at the sound. Stereo drum miking had been born.

Johns continued to use this setup while working with some of the biggest acts of the day, including The Rolling Stones, The Who and The Eagles.

Considerations

- The room plays a big factor in the sound. The larger the room, the closer to the sound of those records it will be.

- The drummer also plays a big factor. A heavy hitting drummer will get much better results than a light hitter.

- Phase between the two "overhead" mics is critical, so try to keep them about the same distance from the snare drum.
- Both of the top mics are panned left and right although not 100% hard left and right and not at exactly the same panning position. Panning will provide the balance of the kit.
- Johns used Neumann U 67s as overheads and an AKG D 12 for the kick.

Placement

Technique #1: This technique involves three directional microphones. The first is a standard overhead placed 4 to 5 feet directly over the snare drum. The second mic is placed looking over the floor tom, but aimed at the snare. The third mic is a standard kick mic (see Figure 8.42).

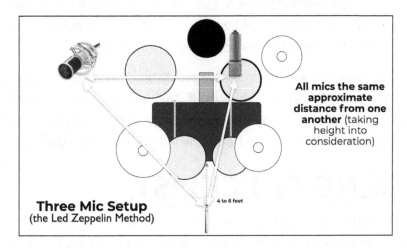

Figure 8.42: A three mic setup using the Led Zeppelin method
© 2023 Bobby Owsinski (source: iStock Photo)

- **Variation:** Add a snare mic and send it to a different channel (Johns often did that as well).

Technique #2: The "When The Levee Breaks" sound was achieved using two Beyer M 160 ribbon microphones hung above the drums with no kick mic. The sound comes from both the spacious hallway at Headley Grange, coupled with lots of compression.

The '80s Drum Sound

Music was changing as the 1970s rolled into the 1980s, and the sounds changed as well. Studios went from spacious and roomy to smaller and drier sounding. The sound of the early '80s during the Disco era was more defined by the gear used than by the microphone placement.

Considerations

- If there ever was one drum kit that defined the sound of the '80s it was the Yamaha Recording Custom with Remo Pinstripe or Evans Hydraulic heads. As the decade moved along the preferred head choice was Remo Ambassador Clear or Emperor Clear.

- The snare was often tuned low so it almost sounded like a tom. Light muffling was used to cut down on the ring.
- The sound of the snare drum was enhanced with either a gated reverb or a non-linear setting from an AMS RMX16 (Program 8 -- NON LIN 2).
- Typical mics chosen were AKG D 12 on kick, Sennheiser 421s on toms, SM57 on snare, and AKG 451s on high-hat and cymbals.

The Reggae Drum Sound

Looking for that Bob Marley sound? Here are some things to try:

- Dead heads (usually old clear pinstripe or Evans hydraulic heads) and boxy-sounding toms are the norm if you want that old-school Lee "Scratch" Perry dub sound. Try a timbale tuned up really high and place it where the floor tom would normally be.
- Wailers drummer Carlton Barrett's snare was a vintage Ludwig 5 ½ x14 with a coated head tuned up until the snare began to choke itself. This gave a timbale-like effect when struck on the edges of the head (slightly ringy but controlled) with the advantages of having a loud and snappy cross rim shot. The tighter the head, the snappier the rim-shot sound.

DRUM MIKING CHECKLIST

Like the foundation of a house, the drums are the foundation of a recording. With a strong foundation, you can build almost anything on it that you or your clients can imagine. A little time and effort spent miking the drums and getting the sound just right can result in a recording that sounds better than you would have ever imagined. That said, here's a drum recording checklist to go over if something just doesn't sound right.

Remember that each situation is different and ultimately the sound depends upon the drums, the drummer, the room, the song, the arrangement, the signal chain, and even the other players. It's not unusual to have at least one of these things out of your control.

- ☐ **Are the mics acoustically in phase?** Make sure that tom mics and room mics are parallel to each other. Make sure that any underneath mics are at a 45° angle to the top mics.
- ☐ **Are the mics electronically in phase?** Make sure that any bottom mics have the phase reversed. Make sure that all the mic cables are wired the same by doing a phase check as described in Chapter 5.
- ☐ **Are the mics at the correct distance from the drum?** If they're too far away they'll pick up too much of the other drums and induced acoustic phase cancellatio. If they're too close the sound will be unbalanced with too much attack or ring.

- ☐ **Are the drum mics pointing at the center of the head?** Pointing at the center of the drum will give you the best balance of attack and fullness.

- ☐ **Are the cymbal mics pointed towards the bell.** If the mic is pointed at the edge of the cymbal, you might hear more air "swishing" than cymbal tone.

- ☐ **Is the high-hat mic pointed at the middle of the hat?** Too much towards the bell will make the sound thicker and duller. Too much towards the edge will make the sound thinner and pick up more air noise.

- ☐ **Are the room mics parallel?** If you're using two room mics instead of a stereo mic to mic the room, make sure that the mics are on the same plane and are exactly parallel to each other. Also make sure that they're on the very edge of the kit looking at the outside edge of the cymbals. There are many other ways to set up room mics, but this is a great place to start.

- ☐ **Does the balance of the mix sound the same as when you're standing in front of the drums?** This is your reference point and what you should be trying to match. You can embellish the sound after you've achieved this.

- ☐ **Have you checked the drummer's headphones?** The cue mix is critical for a drummer, so make sure the headphones are in good working order with at least a basic mix balance. You'll undoubtedly tweak it later.

Remember, take risks, experiment, take notes on what works and what doesn't, be creative, and most of all, have fun! The items on this drum recording checklist are not hard and fast rules, just a starting place. If you try something that's different from what you've read and it sounds good, it is good!

9

MIKING INDIVIDUAL INSTRUMENTS

This chapter provides a variety of miking approaches for individual instruments and ensembles that I've collected over some 30 years from other engineers, producers, mentors, manufacturer's reps, and musicians. (Unfortunately, in most cases I can't remember who showed me what, and in some cases I was sworn to secrecy never to tell.) Every technique works, at least to some degree, but what will work for you depends upon the project, the song, the player, the room, and the signal chain. **Since no two situations are alike, use these approaches as merely a starting point.** Experiment, take what works, and leave the rest.

Because there are a lot of factors that go into recording something well, this section is treated somewhat differently from what you might expect. First of all, unless there is a very specific need to use a particular microphone for an application, I will suggest just the general type of mic (in other words, ribbon, dynamic, condenser) if needed. One of the reasons for this is the fact that not everyone has as wide of a variety of high-end microphones available to them as many of the applications might suggest. Second, the mic itself usually has less to do with the ultimate sound than the placement, room, player, and ultimately, the project itself. Even if all you have are inexpensive mics, the techniques can still work.

ACCORDION

Accordion is a central instrument in zydeco, Cajun, and polka music, although it's also found in indigenous music around the world. There are many different types and sizes of accordion, but the following techniques cover them all.

Just to show that all roads sometimes lead to the same destination when it comes to the type of microphone used, it's interesting to look at some of the mics used on the albums by accordion virtuoso Dick Contino back in the '50s and '60s. On an album engineered at Universal Studios in Chicago by the legendary engineer and equipment designer Bill Putnam, a U 47 was used. On another album tracked by engineer Malcolm Chisholm at United Western Studios in Hollywood, an RCA 77-DX was used. On another Universal/Chicago date, engineer Bernie Clapper used a 251.

Considerations

- Like many instruments, an accordion radiates a different tone in every direction, and each accordion surface produces a distinct timbre. As a result, the tonal balance can be dramatically altered depending upon where a mic is placed.
- When playing with a rhythm section, many accordionists play only with the right hand.
- If button and air noise is a concern, a single dynamic mic (such as an SM57) tends to pick up less of the noise from the instrument than a condenser mic. This is also a good choice for the button-type instruments used in Tejano and Norteño music

Placement

Technique #1: The most basic technique is to place a mic about 2 or 3 feet away from the bellows (the pleated layers that the player moves to push air over the reeds) of the accordion.

- **Variation:** Instead of a single mic, use a stereo mic or a coincident pair. The sound will no longer come from just one point in space, plus it will sound more natural. Experiment with a little more distance than the 3 feet mentioned above when using a stereo pair.

Technique #2: If the player will mostly be playing with only his or her right hand, place a mic on the keyboard side about a foot or so away from the reeds above the keyboard.

Technique #3: Clip a miniature lavaliere mic to the wrist strap of the accordion.

Technique #4: A standard pickup arrangement for a Cajun accordion is an SM57 capsule mounted on a bracket at the bottom of the accordion, facing upward. These are usually four-reed accordions played with all the stops (the buttons that select the reeds that are used for different pitch or colors) selected so the sound is at its fullest.

Technique #5: For internal miking of the reeds, place three lavaliere mics on the treble side (low, middle, high) and two on the bass of a full-sized piano accordion.

AUDIENCE

Audience recording is both the key to and the problem with live recording. It's sometimes difficult to record the audience in a way that captures its true sound. The transient peaks of the audience make it difficult not only to capture well, but also to isolate from the stage mics.

Considerations

- Miking the audience lends itself to using omnidirectional mics, but shotgun mics can be especially useful because they help attenuate the intimate conversations from the crowd that happen around where the mic is placed. In the event that you have neither type of microphone, just make sure that the mics you use are identical models, and don't forget to engage the low-frequency roll-off switch if the mic has one.

- Mic placement outdoors is a lot more difficult because you have nothing to hang microphones from to get enough distance above the audience. For another thing, you don't have the ambiance of the venue to help you out, so you usually have to resort to more microphones as a result. Don't forget the windscreens, because nothing makes a track unusable like wind blasting across the mic's capsule.

- Many engineers are tempted to use stereo recording techniques, such as spaced pairs, X/Y, ORTF, and Blumlein, but these can actually return some poor results when it comes to audience miking. What these setups do is capture the ambiance of the environment and a perfect stereo picture, but your primary concern is just to capture the audience. They're two different beasts and have to be handled that way.

- It's very easy to have the output level of audience microphones overload the mic preamps they're connected to either because of the stage volume of band or the peaks of the audience response. Therefore, it's a good idea to use the attenuator pads and heavily compress or limit them to prevent overload.

Placement

Technique #1: Place a pair of identical mics at about the halfway point between the edge of the stage and the back wall of the venue. Make sure that the mics are placed at least 3 feet above the audience. Start with the microphones facing directly at one another across the audience, as in Figure 9.1, and then aim them both down toward-but not exactly at-the middle of the audience.

TIP: The higher you get the mics over the audience the better the audience will sound. However, if you're in a club with a low ceiling, you're better off with placement closer to the audience, since the reflections from the ceiling can sometimes negatively affect the sound.

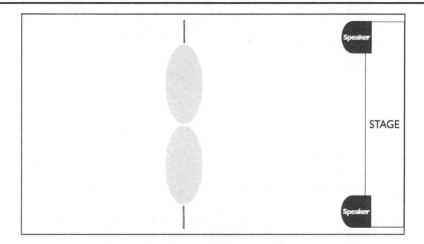

Figure 9.1: Simple audience-miking technique
© 2023 Bobby Owsinski

- **Variation:** If you need only a *mono* audience track, splay the mics off-access, as in Figure 9.2. This configuration can result in a fuller sound in mono but will result in a stereo track that's off balance, since one mic is pointed closer toward the stage and the main sound system than the other.

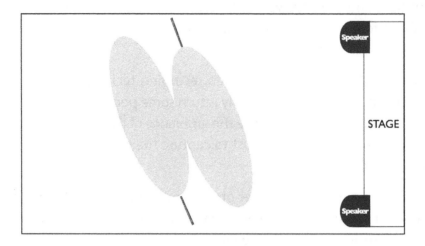

Figure 9.2: Simple audience-miking technique variation
© 2023 Bobby Owsinski

Technique #2: In a club, hang a couple of mics at about the middle of the venue, pointing directly down from the ceiling. This is where omnidirectional mics come in handy. Be sure to hang all mics at the same distance from the stage to get a balanced stereo image.

Technique #3: Place two identical mics directly in front of the stage between the front-of-house speakers and pointed at the middle of the room as in Figure 9.3. The trick is to place the mics so their null points are directed at the house speakers so you can hear them the least and the low frequencies will be at their weakest.

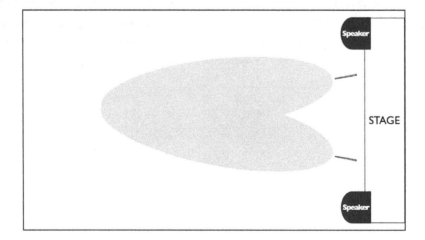

Figure 9.3: Miking the audience from the front of the stage
© 2023 Bobby Owsinski

- **Variation:** Use two mics pointing toward the center and two toward the side. This technique works great when you just can't find anywhere secure to place microphones in the crowd. If there's a balcony, aim the mics at the farthest seat instead of the middle of the room.

Technique #4: Placing mics on the backline of the stage not only will give you a great drum sound (not that you're looking for one), but will also give you a great audience sound. Place them on tall stands on the back of the stage pointing at the back of the singer's head, as in Figure 9.4.

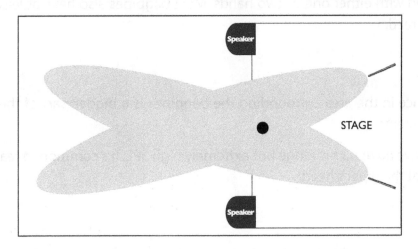

Figure 9.4: Miking the audience from the rear of the stage
© 2023 Bobby Owsinski

- **Variation:** In cases where you're using the #3 or #4 miking configurations, you'll usually need a pair of rear hall mics as well. Face them forward, looking at the stage, but try to place them 6 feet or so from the rear wall or the corner of the venue so they don't pick up any unwanted reflections.

Technique #5: Sometimes the easiest place to put the audience mics is at the front-of-house console, especially in a large venue. Assuming that you're set up in the middle of the audience and not under a balcony or some other obstruction, try placing four mics (preferably shotguns) at the corners of the mixing position, as in Figure 9.5. The front two mics will be aimed toward the front of the house, just in front of the PA stacks. The rear mics will point into the deep house left and right, toward the corners of the venue. This method works great by itself and even better with the addition of front hall or backline mics.

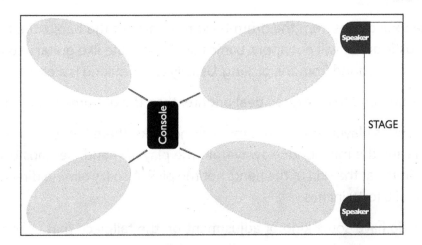

Figure 9.5: Audience miking from the console position
© 2023 Bobby Owsinski

BAGPIPES

Bagpipes use enclosed reeds fed from a constant reservoir of air in the form of a bag. The chanter is the melody pipe played with either one or two hands. Most bagpipes also have at least one drone pipe, which has a single reed.

Considerations

- The ambiance in the area surrounding the bagpipes is a bigger part of the sound than with most instruments.
- Bagpipes have no dynamic range but extremely high SPL. It's common to read levels as high as 108dB SPL at the piper's head.

Placement

Technique #1: Place an omnidirectional mic at least 3 feet above the piper, pointed down.

Technique #2: For stereo, try a pair of baffled omnis (like a Jecklin Disk - see Chapter 6)) or a set of cardioids in ORTF configuration. Place on a high stand or boom 2 to 3 feet above the piper, pointing down.

BANJO

Although the banjo is most identified as a country or bluegrass instrument, its origin is actually with African slaves, who brought the instrument to America. Banjos come in four, five, or six string variations, but they all have the same general sound.

Considerations

- The banjo will tend to sound harsh with a lot of midrange because, after all, it's a banjo, and they sound harsh with a lot of midrange by nature! Try a ribbon or dynamic mic to mellow out the sound a bit.
- Since all the tone comes from the drum-head resonator of the banjo, the techniques used to mike an acoustic guitar will not apply. Banjos don't resonate like guitars, so it's better to try to capture the attack sound from the picking. Usually close-miking is a good technique for this.
- Bluegrass banjos tend to be a good deal brighter than the old-time open-back banjos.
- Because bluegrass players usually use metal fingerpicks, there tends to be a good deal of pick noise. Try placing the microphone away from the player's hand, perhaps below the bridge so that you're aiming at the skin of the hand, not the pick. Also try aiming directly below the hand at a distance of 8 to 10 inches.
- Most banjos do have some kind of adjustment on the tailpiece that changes the amount of downward pressure the bridge puts on the head. This will have some effect on the attack and tone of the instrument.

- Don't neglect your microphone preamp. A bluegrass banjo is about as good a torture test of a preamp as there is. The better your preamp, the less trouble you'll have getting the sound.

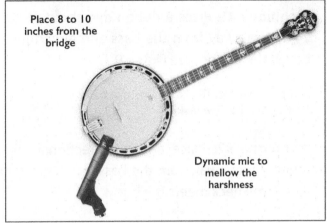

Placement

Technique #1: Place a mic 8 to 10 inches away from the front of the bridge, which captures the sound of the whole body (see Figure 9.6).

Figure 9.6: Basic banjo-miking technique
© 2023 Bobby Owsinski (Source: iStock Photo)

Technique #2: Place a mic facing down about 2 to 3 feet above the banjo and a foot in front of the player's head.

Technique #3: Place a mic 6 to 8 inches from the base of the picker's hand or just above, depending on the instrument and the picker.

Technique #4: Place two mics from 6 to 18 inches from the front of the banjo. Point one mic in the proximity of where the neck meets the body (or even a little higher up the neck), and then point the other mic in the proximity of the center of the resonator head, where the bridge and the player's picking hand are. Experiment with the mic pointed at the head, as different angles and slightly different positions can produce quite different sounds.

Technique #5: To reduce the noises that occur from the picking hand brushing against the head, clip an omni lavaliere mic to the strap down by the neck.

BASS (ACOUSTIC STRING OR UPRIGHT)

Acoustic string bass is one of the hardest instruments to capture well, usually because it's being played in a live setting, such as a jazz trio, and placed very close to other instruments. If you get it to sound good, you might have a lot of drums and piano bleeding into the track as well, which then limits your control when mixing. It doesn't have to be this way though, as there are a number of tried and true methods that work great and give you the isolation needed.

Considerations

- Positioning is everything when recording string bass. Close-miking the F-hole usually makes the sound muddy, with no definition.

- Perhaps more than any other instrument, the bass needs space to really sound right. Close-miking it can kill it dead if not done with care.

Placement

Technique #1: Place a ribbon mic about 18 to 24 inches away from the bass and aimed just below the bridge (see Figure 9.7).

Figure 9.7: Royer R-121 on upright bass
© 2023 Bobby Owsinski

Technique #2: Place an omnidirectional mic about 2 feet away from the bass, pointed near where the neck meets the body.

Technique #3: Place a mic about 18 inches away from the bass, about halfway between the bridge and the end of the fingerboard.

- **Variation**: Use a combination of a condenser mic placed as above and an SM57 wrapped in foam under the fingerboard overhang.

Technique #4: Place a large-diaphragm condenser aimed at the strings from about 8 to 12 inches above the strings but below the bridge (see Figure 9.8).

Figure 9.8: Soundelux U95 on upright bass
© 2023 Bobby Owsinski

Technique #5: Place a ribbon mic *behind* the bass at about bridge height to capture the warmth of the wood.

Technique #6: Place a small-diaphragm condenser about even with the end of the fingerboard, pointed down halfway between the fingerboard and the bridge and about 18 inches away.

Technique #7: Place a dynamic mic wrapped in a piece of foam and nestled in the tailpiece (pointing towards the bridge).

Technique #8: Here's a trick from Paul Langosch, bassist extraordinaire with Tony Bennett for many years. Wrap a small-diaphragm condenser in foam and wedge it between the A and D strings, aimed between the feet of the bridge. You may need to use the 10dB pad on the mic or preamp if it's a loud bass.

Technique #9: You can see in pictures from Rudy Van Gelder (engineer for all the famous Blue Note recordings of the '50s and '60s) sessions that he was fond of using an RCA 77-DX placed near the floor, angled up toward the bridge at about a 45-degree angle. This can greatly reduce leakage from other instruments, since the null of the mic is pointed outward.

Technique #10: Place a ribbon mic 4 inches above the right hand on an upright. Orient it more toward higher strings.

BASS (ELECTRIC)

Just like the acoustic bass (and just about all instruments for that matter), a great bass sound is dependent upon the bass, the player, the amp, and the room. The player has to be able to achieve the tone you're trying to record with either a pick or fingers first and foremost.

Considerations

- Although using only a DI can sound good for bass, using an amp (or both together) can really make it easier to dial in a great sound. However, many times the frequency band of the amp can mask the frequency bands of other instruments, such as guitars.
- Always check the phase relationship between the amp and DI to make sure there's no cancellation of the low end. Flip the polarity switch to the position that has the most bottom.
- If you're recording into a DAW, align the bass amp track with the DI track so that they're more in phase (you really have to zoom in tight for this).
- There's no rule that says you have to use both the DI and amp tracks together, so don't hesitate to use just one of them if it sounds best in the mix.

Placement

Technique #1: The simplest and cleanest way to record an electric bass is with a direct box. Be careful which one you choose, because some will not capture the low fundamental of the bass. Active DIs do a better job of this than passive, although some passive boxes (such as the ones made by Radial) do an excellent job because of the large Jensen transformer used in the circuit.

Technique #2: When miking a bass amp, place a large-diaphragm dynamic, like a D 112, an RE20, or a BETA 52, a little off center and a couple of inches away from the cone of the best-sounding speaker in the bass cabinet (if it has more than one). Depending on the sound that fits the track best, mix with a DI track. The sound will change substantially depending upon the balance between the DI and the miked amplifier.

Figure 9.9: Shure BETA 52 and SM57 on a bass cabinet
© 2023 Bobby Owsinski

- **Variation:** Add an SM57 to the large-diaphragm mic sound for more bite, but be careful not to use too much, or you'll lose the main body of the sound (see Figure 9.9).

Technique #3: For a metal bass sound, try splitting the bass signal into the normal bass amp and also into a guitar amp that's set to distort. Make the sound a little dirtier than you feel is appropriate, as it may help it to sit better in the mix. Adjust the EQ of the guitar amp to taste and then add the DI to taste.

- **Variation:** An overdrive or saturation plugin can add the same flavor during mixing.

Technique #4: Raise the speaker cabinet about 3 feet off the ground to decouple it and decrease the reflections from the floor. Mike with a ribbon mic or a large-diaphragm condenser from 3 to 6 inches away from the cone.

Technique #5: To find the sweet spot of the bass amp growl, have someone move the mic across the cone and in and out from the cone until you hear what you're looking for. Nearer to the edge of the speaker will give you more boom, while nearer to the cone will give you more color. Somewhere in between you'll find the sweet spot that will best fit in the mix. Don't worry if the mic ends up in a place that looks wrong.

Technique #6: For that Paul McCartney Beatles sound, place the amp in the middle of the studio (it helps if you have a large room) and place a large-diaphragm condenser 2 to 10 feet away. An AKG C 12 was usually used for this, but it was sometimes augmented with an STC 4038 for an additional darker sound.

- **Variation:** Another McCartney trick was to use a guitar amp with a 2x12 cabinet for bass with the mic placed about a foot away and aimed between the speakers. Many Beatles songs from 1966 to 1968 used this setup.

BASSOON

The bassoon is a relatively modern instrument, eventually evolving into its present form after its creation in the 1800s. Due to its complicated fingering and reed difficulties, it's a very demanding instrument to learn and to play.

Considerations

- As with all woodwinds, the sound from a bassoon emanates along the entire body of the instrument with some coming from the top.

Placement

Technique #1: Place the mic in front of the instrument at the player's eye level, about 3 to 4 feet away.

Technique #2: Place a mic at bell height about 45 degrees from the player's right side of the instrument.

Technique #3: Place the mic at least a couple feet above and a couple feet in front of the place that sounds best, which often depends on the room.

Technique #4: There are two common places to mount a pickup on the bassoon: the F# trill key hole on the wing and/or the bocal (the thin tube that goes from the reed to the body of the bassoon). That said, most bassoonists will not let you clip anything to their bocal because these are typically very expensive and good ones are hard to come by.

Technique #5: Try three microphones. Place a mic over the top of the instrument, another near the middle near the bocal, and one more near the bottom of the instrument (but remember to check the phase between them).

BONGOS

While some people think of bongos as a remnant from the old beatnik days, it's surprising how often they're used to add movement to a track.

Considerations

- Bongos are made up of two drums, the macho (smaller drum) and the hembra (larger drum).
- Tuning can be anywhere between a fourth and an octave between drums, depending upon the music.
- Always detune the smaller drum after use, since the head will either stretch or break because it's so tightly stretched if it's tuned correctly.

Placement

Technique #1: Place a mic from 18 to 24 inches directly above and slightly favoring the smaller drum. If pointed directly in the middle of the two, the larger drum will usually sound louder.

- **Variation:** Sometimes a dynamic mic, such as an SM57, helps the bongos stand out in the mix due to its midrange emphasis.

Technique #2: If the player is sitting, try miking the bongos from underneath. Use a dynamic mic at a distance of at least 6 inches, again slightly aiming toward the smaller drum.

BOUZOUKI

Although one immediately thinks of Greek music upon mention of the bouzouki, it can be found in music from all over Europe, especially in the bands from Ireland, Scandinavia, France, and Turkey.

Considerations

- The bouzouki has almost no low end and can sound very thin and metallic if miked carelessly. The desired tone is usually quite rich in lower mids with a clear but clean treble.
- As with most other acoustic instruments, the quality of the player and instrument are crucial.

- Many older bouzoukis are "three course" (three sets of two strings), while more modern instruments are "four course" (four sets of two strings).
- The Irish bouzouki has four courses and a flatter back.

Placement

Technique #1: The same mics and techniques used for acoustic guitar can work for bouzouki. To start, place a small-diaphragm condenser mic 8 to 12 inches away, aimed at a spot between the bridge and the sound hole. If more low end is needed, move the mic closer to the sound hole.

- **Variation:** To augment the low end, tape a lavaliere mic near the sound hole.

BRASS

Since many of the miking techniques are the same for different brass instruments, we'll consider them as a family, and I'll point out any differences for placement with specific instruments.

Considerations

- With the mic aimed directly at the bell from a close distance, every bit of spit, excess tongue noise, air leak, and all the other nasties that every brass player occasionally produces is much more apparent. Either moving the mic back a little or pointing it a little off axis of the bell can hide the majority of these unwanted extraneous noises without compromising the natural tonal color of the instrument too much.
- What are the differences between a trumpet, cornet, and flugelhorn? Trumpet is one-third flared tubing and two-thirds straight. Cornet is half and half. Flugelhorn is two-thirds flared tubing and one-third straight. A flugelhorn is really a soprano tuba.
- If the brass instrument itself is shrill, try placing a sock loosely in the bell (which the player probably won't like), lowering the mics, or pointing them slightly off-axis.
- The recording environment plays a big role in the sound as well. Warm sounding rooms with lots of porous wood tend to help the sound as well as help the players hear themselves.

Placement

Brass instruments require placement of the mic some distance from the instrument in order to sound natural, as the sound needs space to develop. Most of the following placement revolves around different ways of obtaining that distance.

Trumpet

Technique #1: Place the mic about 4 feet away but directly in front of the instrument.

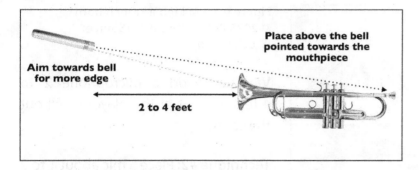

- **Variation**: Place the mic 3 to 4 feet away, but above the bell and aimed toward the mouthpiece (see Figure 9.10).

Figure 9.10: Miking a trumpet
© 2023 Bobby Owsinski (Source: iStock Photo)

Technique #2: For a brass section, place the players in a circle around a Blumlein pair (crossed figure 8's - see Figure 9.11). Be sure that each player is on the lobe of the pattern and not in the null point. Balance the section by moving the softer horns closer to the mic and the louder ones farther away.

Figure 9.11: Horn players in a circle
© 2023 Bobby Owsinski (Source: iStock Photo)

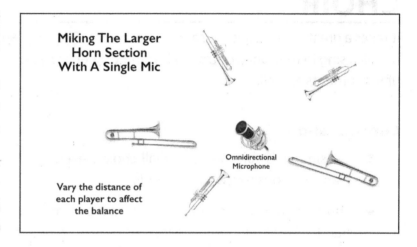

- **Variation 1**: If a stereo spread is not needed, try something as simple as an omni condenser in the middle of a circle of players.
- **Variation 2**: Try hanging an omni condenser about 4 feet directly over the horn group as an additional mic.

Trombone

Technique #1: If you are going for a more mellow jazz type of sound, place the mic about 20 to 30 degrees off axis of the bell at a distance of about 2 feet. If you want a more aggressive sound that will cut through a dense mix, mike the horn directly in front of the bell and move the mic in to about 12 to 18 inches.

Technique #2: Ribbon mics are great for trombones at a distance of about 6 to 12 inches from the bell.

Tuba

Technique #1: Position the mic about 2 feet over the top of the bell at about 15 degrees off axis of the center. If it sounds too "blatty," aim the mic more off axis.

French Horn

Technique #1: Place a mic from 2 to 4 feet from the bell at a height that matches the top edge of the bell but aimed toward the center (see Figure 9.12).

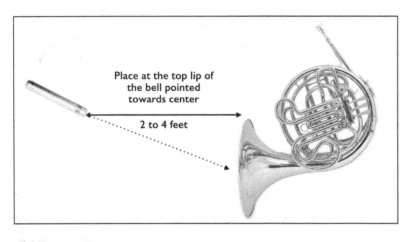

Figure 9.12: Miking a French horn
© 2023 Bobby Owsinski (Source: iStock Photo)

Variation: Add a microphone 2 to 4 feet in front of the player to fill out the sound.

Technique #2: Place a mic about 1 to 2 feet over the player's head and pointed straight down.

CHOIR

It takes a great choir to get a great choir recording. Singers need to not only blend well with each other, but also sing in tune and control their volume. Keep in mind that these techniques apply to ensembles of instruments as well.

Considerations

- Ten singers is considered a small choir. Eleven to 25 is considered a medium-size choir. More than 25 is considered a large choir.
- Choirs are always arranged in SATB sections, meaning sopranos, altos, tenors, and basses. The standard formation for classical choirs is from high to low with the highest soprano standing on the far left and the lowest bass on the far right, although other arrangements are also used.

Placement

Technique #1: Place an ORTF stereo pair about 3 feet above the conductor and aimed at the center of the choir as a starting point (see Figure 9.13). A coincident or near-coincident stereo-miking technique will likely be far preferable to a widely spaced omni-pair technique. This is because you generally want to hear the interplay of vocal lines within the choir as they move about the sections, and spaced omnis won't necessarily provide that.

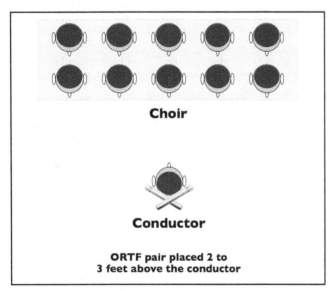

Figure 9.13: An ORTF stereo pair recording a choir from the conductor's position
© 2023 Bobby Owsinski

Technique #2: For a big but somewhat unfocused choral sound, try a pair of spaced omnis.

Technique #3: Try legendary engineer Bruce Swedien's method, which is a Blumlein pair (crossed figure 8's) with the choir placed in a circle around the mics.

Technique #4: Record the choir in sections to have more control over the balance during the mix. Place a mic 6 to 8 feet away from each section. The farther from each section the mics are placed, the better the blend will be, but the more the other sections will bleed into the section mic (which defeats the purpose). You also won't have a true stereo recording, which might not give you that "you are there" listening experience.

- **Variation:** Augment any of the above techniques with spot mics to fill in any deficiencies in the stereo miking.

CLARINET

Although more closely identified with Dixieland, classical, and swing music, the clarinet has been used prominently on a few rock hits, such as the Beatles "When I'm Sixty-Four" and Sly and the Family Stone's "Dance to the Music."

Considerations

- Clarinets have been made from a variety of materials, including wood, plastic, hard rubber, metal, and ivory, but the vast majority of the quality instruments are made from African hardwood.
- It's difficult to close-mike a clarinet effectively with just one mic. If you place the mic at the top of the instrument, the bottom notes are weak. Unlike the sax, most of the sound comes straight out of the bell at the lowest overblown note.
- On the other hand, miking the bell results in a weakness in the bridge notes between the fundamental and first overblown range. Clarinetists spend years working on their tone through this area, so you need to support it when miking.

Placement

Technique #1: Place a mic pointing down at the small "A" key, with a second mic placed just off the bell.

Technique #2: If only one mic is available, use an omni placed about 2 feet away from the bell and 2 feet above the instrument, pointed down.

CLAVES

Claves are very important in Afro-Cuban music, since they're frequently used to play repeating rhythms. Most modern claves are no longer made of wood and instead use plastic or fiberglass because of its durability. The clave is also the name of a basic rhythmic pattern.

Considerations

- Claves are very loud and sometimes require baffling to control the room reflections.
- There's not much body or decay from a clave, so the room in which you're recording is very important.

Placement

Technique #1: Place a dynamic mic about 10 feet away and at about shoulder height of the player, but pointed toward the claves.

CLAPS

See the "Hand Claps" section.

CONGA

Congas come from Cuba, although their origin can be traced back to Africa. Modern congas have a staved wooden or fiberglass shell with a screw-tensioned drum head.

Considerations

- For congas, a hard floor in a fairly large room is essential to getting a good, natural sound. A hardwood floor is the best, but tile or linoleum will do as well.
- The larger the room, the better.
- Congas sometimes sound better when placed directly on the floor than they do on a stand.

- The traditional tuning of the conga was much lower than what's used today on most records.

Placement

Technique #1: Place a small-diaphragm condenser or dynamic mic about 1 to 2 inches in from the outer rim, and hovering about 12 inches above each drum (see Figure 9.14).

Figure 9.14: AKG 452 over the conga
© 2023 Bobby Owsinski

- **Variation 1:** Add a dynamic mic below the congas, sitting on the floor and facing up between the drums to fill out the sound.

- **Variation 2:** Add an additional room mic positioned about 6 feet away from the drums and 6 feet high.

Technique #2: Place the mics so they're a few inches below the rim, under the congas and angled up and aiming at the player's eyes (see Figure 9.15).

Figure 9.15: AKG 452 under the conga looking up
© 2023 Bobby Owsinski

Technique #3: When using only a single mic, place it 12 to 18 inches between the drums, but slightly aimed toward the one that's higher tuned.

COWBELL

The cowbell actually had its origins as a way for a herdsman to identify his or her freely roaming animals. Today it's a widely used percussion element frequently used in many genres of music, but a mainstay in salsa music.

Considerations

- Cowbells project a high-frequency transient from the closed end as well as a fundamental frequency from the open end.
- The high end of the cowbell easily cuts through the mix, but the low frequencies sometimes get lost.
- Cowbell players sometimes move around when they play, so be sure that the mic is far enough away so they won't move off-mic.

Placement

Technique #1: Place a mic about 2 feet in front of the cowbell but about a foot above it and angled downward (see Figure 9.16).

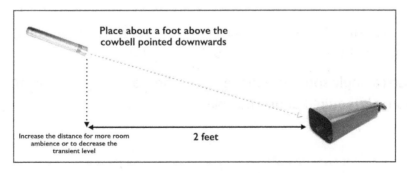

Figure 9.16: Miking a cowbell
© 2023 Bobby Owsinski (Source: iStock Photo)

Technique #2: Place a mic about 12 to 18 inches away, below the mouth of the bell and angled upward.

CROWD

See the "Audience" section.

DIDGERIDOO

The didgeridoo is a wind instrument used by the indigenous Aboriginals in Australia. The traditional instrument is made from dried bamboo or eucalyptus and has a distinctive single low pitch.

Considerations

- Be aware that the instrument generates a lot of subsonic frequencies, so inserting a high-pass filter in the signal path might prove useful.
- The Didjeribone is a modern, professional version of this ancient instrument. It has the advantage of being able to slide between 10 keys. See didjeribone.net for more info.

Placement

Technique #1: Place a dynamic or a small-diaphragm condenser mic about 4 to 6 inches from the bell of the didj. This is the ideal distance in that it gives the best balance of low end and clarity.

Technique #2: Mount a small clip-on omni condenser on the end of the didj's bell.

Technique #3: Charlie McMahon's Facebass is a seismic sensor that records sounds inside the mouth and is an alternative to using microphones for recording a didj.

DJEMBE

A djembe is a skin-covered hand drum from West Africa that's played with the bare hands. Because of its shape, density of wood, and the head material, it's capable of a wide variety of sounds, especially a very deep bass note.

Considerations

- The heads on most djembes are fairly wide (14 inches or so), but some of the bass sound comes from the bottom of the instrument and not from the head.
- There really isn't a single spot close to the drum where a mic can capture the full djembe sound, so some distance is required during mic placement.

Placement

Technique #1: In a good-sounding room, the drum should be miked from 6 to 10 feet away (see Figure 9.17).

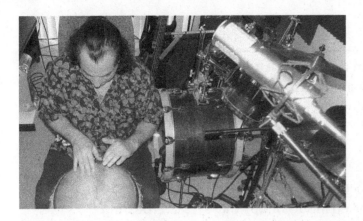

Figure 9.17: Mic placed over a djembe
© 2023 Bobby Owsinski

Technique #2: For close-miking, place a single mic 4 to 6 inches above the rim and angled across the drum head.

Technique #3: To capture the extra low frequencies, place a mic underneath the drum, aimed at the opening.

DOBRO (RESONATOR GUITAR)

Dobro is actually a brand name of a resonator guitar, although the name has become synonymous with the instrument. National is the other company noted for its resonators. A resonator guitar is the same as an acoustic guitar except that a resonator is put in place of a sound hole, which was an attempt to acoustically amplify the guitar. There are two types: round necks, which are played like a normal guitar, and square necks, which are played on their backs on the player's lap, facing up.

Considerations

- The sound of the Dobro really depends upon who's playing more than most other instruments.
- The sweet spot is usually between the treble-side screen and the coverplate, toward the bottom of the instrument.
- The screen area has mostly low end, and the sound gets brighter as you move toward the coverplate.

Placement

Technique #1: Place a mic about 4 to 6 inches away from the screen area and a second mic at the coverplate, aimed at the screen area (see Figure 9.18).

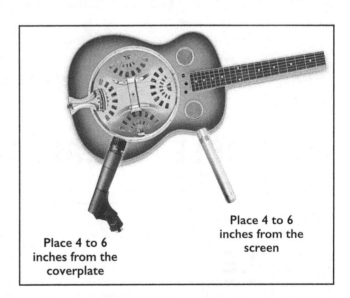

Figure 9.18: Basic Dobro miking
© 2023 Bobby Owsinski (Source: iStock Photo)

Technique #2: Place a mic about 6 inches off the resonator and another about 6 inches off the treble side hole.

DULCIMER

There are two kinds of dulcimer. The Appalachian dulcimer is a fretted instrument with drone strings, while the hammered dulcimer uses small mallet hammers to strike each set of strings. Like many other stringed instruments, the hammered dulcimer uses sets of strings to increase the volume level.

Considerations

- The sound of the pick (or the quill, as it's more traditionally called) is very much part of the sound of a dulcimer, but should not be miked in such a way that's it's overbearing.

- Try moving the mics a little closer to the noter (the left hand) to capture some of the resonant sound without getting as much of the pick.

- There are sound holes both near the bridge and the headstock that produce some of the resonant sound. Unlike acoustic guitars, mountain dulcimers generally do not get boomy at the sound hole.

- The key is to get some distance (at least 2 to 4 feet) so that the low-frequency "thump" sounds that frequently occur aren't recorded.

- It helps if the player knows that he or she doesn't have to play the instrument as hard as might be required in a live gig. If a lighter touch is used, the longer the tuning lasts.

Placement

Technique #1: Place a pair of omnis spaced about 1 to 2 feet apart and about 2 to 3 feet above the instrument.

Technique #2: Place an XY configuration of small-diaphragm condensers at about 3 feet away from the center of the instrument.

ENGLISH HORN (COR ANGLAIS)

See Oboe

ENSEMBLE (VOCAL, SAXOPHONE, OR OTHER)

See the "Choir" section.

FIDDLE

Basically, the only difference between a fiddle and a violin is the type of music that's being played. Fiddle music the world over is mainly dance music, while violins are more associated with classical and orchestral music.

Considerations

- Put a finger in your ear and walk around the instrument, listening with the other ear. You will find the sound changes dramatically with position because the radiation pattern is so uneven, and it's different with every instrument. Find the place where you like the sound and place the mic there.
- Place the mic farther back from the fiddle than you think you need to be, because the mic needs to be far enough away so that the sound can project from the instrument.
- If the fiddle sounds harsh, try pointing the mic slightly off axis, as this will attenuate some of upper mid-frequencies.
- Bow noise and grit are sometimes referred to as rosin noise. However, rosin is the material that's put on the bow hair to make it grab the string better, and it doesn't produce any noise of its own.

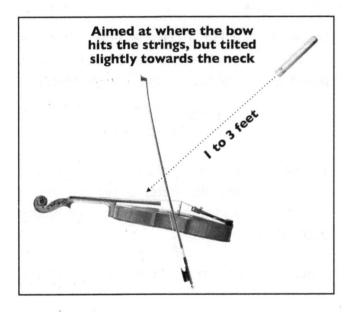

Placement

Technique #1: Place a mic slightly above and about 6 feet in front of the fiddle to capture the full body of the sound without catching too much of the scratchy bow effects (see Figure 9.19).

Figure 9.19: Miking a fiddle or solo violin
© 2023 Bobby Owsinski (Source: iStock Photo)

Technique #2: Place a mic about 1 to 2 feet over the instrument pointed at where the bow hits the strings but tilted a bit toward the neck.

Technique #3: Place a ribbon mic over the player's left shoulder for close-miking.

- **Variation:** Position an additional mic underneath the violin in addition to the top mic.

Technique #4: Position the mic behind the violinist so that the violinist's head and body are partially obstructing the direct path between the mic and the instrument. This is a great way of reducing the ratio of direct to ambient sound without pulling the mic far away from the instrument.

Technique #5: Place the violinist in the corner of the room with the mic about 3 feet away and 2 feet above the instrument.

FLUTE

Flute is one of the least demanding instruments to record accurately. Its pure tone is easily captured by most microphones. The words "transient" and "flute" almost never appear together, except when the music calls for accents.

Considerations

- The higher notes will be closer to the flute blow hole, the lower notes spread more toward the bottom of the instrument.
- Miking too close will pick up a lot of key clicking.
- The flute side (normally the right side) of the player can have more harmonic coloration then the opposite side. The flute side also has more key noise, while the opposite side may have more mouth sounds.
- If you mic near the blow hole, you'll end up with an airy sound. If you mic farther down the instrument, you'll end up with a smoother, not-as-bright sound.
- *Do not record from the open end of the flute.* It doesn't sound like a flute there.

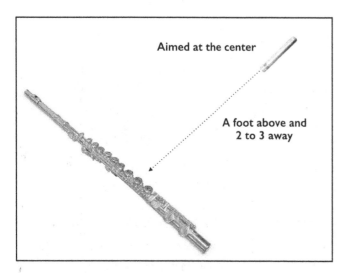

Placement

Technique #1: For a classical flute, place the mic a foot above the flute and 2 to 3 feet away, pointing down at the center (see Figure 9.20).

Figure 9.20: Miking a flute
© 2023 Bobby Owsinski (Source: iStock Photo)

Technique #2: Place the microphone several feet away from the flute, above the flute's embouchure and somewhat off to the side (try both sides).

Technique #3: For jazz flute, place the mic about 6 inches from the mouthpiece to catch all the breath sounds.

FRENCH HORN

See the "Brass" section.

GUITAR (ACOUSTIC)

Acoustic steel-stringed guitar varies so much from instrument to instrument that one miking style definitely won't work on everything. Luckily, there are a lot of different techniques available, and while all the others might let you down, one is sure to work.

Considerations

- Acoustic guitars come in three body styles that you'll encounter 90% of the time: Dreadnought, OM (Orchestra Model), and Parlor.

- They also come in a variety of sizes that include small body (Parlor, 0 and 00), medium body (000, OM and Grand Concert) and large body (Dreadnaughts, Grand Auditorium, and Jumbo).

- Smaller guitars sound warmer with focused mid-range, while large-bodied ones are louder and have more bass. Generally, smaller body acoustic guitars fit better in most songs.

- Generally speaking, the least desirable sound from an acoustic guitar comes from close-miking the sound hole. The sound is much more tonally balanced in the vicinity of the bridge or at the joint of the neck and the body.

- Placing the mic closer than a foot away from the instrument is going to result in a big proximity boost of the low end if you're using a directional mic. Either switch to an omni or back the mic off a bit.

- Since many of the successful miking methods utilize two or more microphones, be sure to listen in mono to check the phase.

- If you have a thin-sounding instrument, a ribbon mic placed 6 inches from the body tends to make it sound a bit heftier.

Placement

Technique #1: Place a mic about 8 inches away from and pointing at where the fretboard meets the body (see Figure 9.21).

Figure 9.21: Miking an acoustic guitar
© 2023 Bobby Owsinski

- **Variation:** Position a second mic pointing at the body, about halfway between the bridge and the end of the guitar, at a distance of 10 inches. This should add body to the sound, and when the two mics are printed to different channels and panned apart, it can sound spacious and lush.

Technique #2: Place an omni mic a foot away from the sound hole. There is no increased low end due to proximity effect because it's an omnidirectional pattern. Move closer or farther away to affect the tonal balance of the guitar, and balance the direct sound against the room sound.

Technique #3: Use a stereo mic about 8 to 12 inches away, with one capsule aimed toward the bridge and the other aimed toward the headstock.

- **Variation:** Try two small-diaphragm condensers in an X/Y configuration. Aim one at the body below the bridge and the other at about the 12th fret.

Technique #4: Place a dynamic mic aimed at the body of the guitar and a condenser over the guitarist's shoulder at about ear height and roughly even with the front edge of the guitar, pointing at the neck. You get two different tones that can be combined in different ways in the mix depending on what the song calls for (see Figure 9.22).

Figure 9.22: AKG C 452 miking the acoustic guitar over the shoulder
© 2023 Bobby Owsinski

Technique #5: One of the best ways to record vocals and acoustic guitar at the same time is with two figure-8 patterned mics. Aim one at the guitar and make sure the null side is pointing toward the vocalist's mouth, and then take another one for the vocal and make sure its null is pointing toward the acoustic.

GUITAR (ELECTRIC)

There's so much more about getting a great electric guitar sound that goes beyond just miking an amplifier speaker that an entire book was written on it - *The Ultimate Guitar Tone Handbook* (Alfred Publishing). That book covers all the things about guitar tone that can't be included here, but what you'll find in the following sections will help you capture a great guitar sound in the majority of situations that you'll run in to.

Considerations

- An amplifier or speaker cabinet usually sounds better if it's raised up off the ground so that the reflections from the floor don't couple or cancel the direct sound from the speakers. Raise it into the air by putting it on a road case, table, or chair.

- You can get different tones by simply moving the mic more toward the speaker's dust cap or toward the surround (the edge of the speaker where it meets the metal basket). Different angles, different mics, and different distances from the cabinet will all alter the tonal quality.

- The guitarist's signal chain can be a huge help or a big hindrance. You'll get a warmer yet aggressive guitar sound by *decreasing* the amount of distortion that might be coming from the player's pedals, then turning up the amp's volume instead to obtain the desired sustain or distortion from the amp and speaker.

- Typically it's best for a player new to the studio to keep the signal chain on the simple side without lots of processing happening before the amp. That being said, some effects are integral to a player's sound.

- So much of a guitarist's tone comes from the fingers instead of the amp. Great players can coax great sounds from mediocre equipment, but mediocre players can't necessarily get great sounds from great equipment.

- On the typical 4x12 speaker cabinet (such as the standard Marshall 1960 model), the sound from the four speakers usually combines at a distance of 18 to 24 inches from the cabinet center.

- When doubling or adding more guitars, it's best to have a variety of instruments and amplifiers available. Two guitars (a Les Paul and a Strat, for instance) and two amplifiers (a Fender and a Marshall is the classic combination) combined with different pickup settings will allow a multitude of guitar tracks to more effectively live together in the mix.

- In an odd paradox, smaller amps and speakers tend to sound bigger than large amps/speakers when recording.

- To find the sweet spot on a speaker, put on headphones and listen to the amplifier hiss as you move the mic around on the speaker. You can also insert pink noise into the input if the sound of the amp at idle isn't loud enough. If you can remember what the hiss sounded like when you had a good guitar sound, then that's a good place to start.

- Ask the guitar player to turn the tone control on the guitar back a touch. This warms things up and makes it sound a little bigger, especially if you're layering three or four guitars on top of one another.

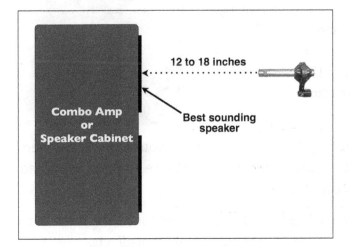

Placement

Technique #1: The old-school setup as done in the '50s and '60s: Place the mic about a foot to 18 inches away from the center of the best speaker in the cabinet (see Figure 9.23). Move toward the edge of the cone if it sounds too bright.

Figure 9.23: The classic method of miking the guitar cabinet
© 2023 Bobby Owsinski

Technique #2: The classic modern setup: Place an SM57 about 1 inch away from the best-sounding speaker in the cabinet. Place the mic about three quarters of the way between the edge of the speaker and the voice coil (away from the voice coil); then move it toward the voice coil for more high end or move it toward the edge of the speaker for more body. Make sure that the mic does not touch the speaker cone during the loudest passages (and longest speaker excursion) played (see Figure 9.24).

Figure 9.24: The modern standard amp-miking technique using a Shure SM57
© 2023 Bobby Owsinski

- **Variation 1:** Add a ribbon mic right next to the SM57, or on a different speaker in the cabinet. A standard setup today is the SM57/Royer R121 combination. The 57 provides the midrange edge while the ribbon mic provides the body.

- **Variation 2**: Place the SM57 as above and then add a Sennheiser MD 421 at the same position to the right of the 57, at a 45-degree angle pointing toward the voice coil. Many sounds can be achieved from this setup by summing the mics at different levels or by flipping the phase on one (see Figure 9.25).

Figure 9.25: A Shure SM57 with a Sennheiser MD 421 placed at a 45-degree angle on a guitar cabinet
© 2023 Bobby Owsinski

- **Variation 2:** Along with any of the above methods, place a ribbon mic 2 inches off one of the rear corners of a Marshall cabinet in order to capture the low end of the cabinet. This only works with Marshall 1960 cabinets due to the wood used and the construction that's unique to Marshall (but it works really well!). See Figure 9.26.

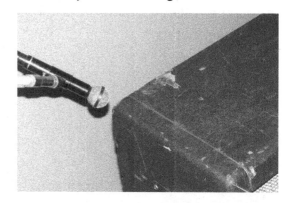

Figure 9.26: Beyer M 160 placed on the corner of a Marshall 1960 cabinet
© 2023 Bobby Owsinski

- **Variation 3:** Add ambiance to any of the above setups by placing a figure-8 mic in the same position as the first mic but at a right angle to the amp to create a side channel for later decoding with an M-S decoder.

- **Variation 4:** Add an additional mic to the above at the spot where the sound of the speakers converges, 18 to 24 inches away from a multi-speaker cabinet. This distance might be increased to as much as 6 feet depending upon the size and sound of the room.

- **Variation 5:** For more ambiance, add a third mic facing a hard wall in the room. The three mics can be mixed together in various proportions to create many different tonal effects.

Technique #3: After finding the correct position as shown above, bundle an SM57, a 421, and a Beyer M 160 (or other ribbon mic, such as a Royer R-121) together. All three mics are aimed directly at the speaker. Add together to taste. The 57 will provide the bite, the 421 the mids, and the 160 the body (see Figure 9.27).

Figure 9.27: A Sennheiser MD 421, a Royer R-121, and a Shure SM57 placed in a cluster on a guitar cabinet
© 2023 Bobby Owsinski

Technique #4: Position a single mic 10 to 20 inches from the cabinet, dead center to all four speakers or, if a Marshall cabinet, aiming at the logo plate.

Technique #5: With an open-back amplifier (such as a typical Fender), place a mic in the rear of the amp, off center from one of the speakers. If used in conjunction with a mic in the front of the amp, try flipping the phase of the rear mic and use the position that sounds best.

Technique #6: Though a little dated, the archetype for that "LA clean rhythm guitar sound" popular in the '80s is a DI'ed guitar, compressed by at least 6dB, with a 25ms delay on the left side and a 50-ms delay on the right.

Technique #7: Use a tiny battery-powered amp like one of the Mini-Marshalls or a Fender Mini-Twin. Close-mike the speaker. The result can be surprisingly large-sounding.

Technique #8: Another one that usually only works well with clean guitars is to tune an acoustic to the key of the song using an open tuning, then place it on a stand near the amp. The amp will make the strings resonate. Position a mic on the body pointing toward the sound hole. The sympathetic vibrations of the strings give you an instant tuned reverb chamber.

Technique #9: For a clean, resonant sound, place the amplifier or speaker cabinet under a piano. Put a weight on the sustain pedal and have someone hold down every key on the piano tuned to the song so the piano strings ring out sympathetically. For example, if the song is in the key of E (major or minor), then hold down all E and B keys. Mike the piano as in the "Piano" section.

GUITAR (NYLON OR GUT STRING)

Even though a nylon-string guitar is still a guitar, the way it's made and the way it's played bring up a whole different set of challenges.

Considerations

- Everything that applies to acoustic guitar also applies to a nylon-string guitar, plus the following:
- With a nylon-string guitar, much of the sound is projected toward the floor from the left side of the guitar (as you're looking at it) if the guitarist sits in the classical position with a footrest.
- Consider putting a small carpet of some sort under the mics to minimize floor reflections.

Placement

Technique #1: If the guitarist is right-handed, place a mic 2 to 3 feet to his or her right and close to the floor, pointing up towards the guitar. Place a second mic 2 to 3 feet away, just a little up the neck from the sound hole on the guitarist's left side. Note that if the mics are placed too close to the instrument, the fret noise will be emphasized. This should work in stereo or mono, provided that the phase relationship between the mics is correct.

Technique #2: Place a small-diaphragm condenser about a foot to the left of the player's left ear, looking down on the 12th fret. Add a large-diaphragm condenser about 12 inches from the strap peg.

Technique #3: Place a small-diaphragm condenser mic 6 to 8 inches from the sound hole but pointed either at the bridge or where the neck meets the body. Move closer to the sound hole for more low end.

Technique #4: Place two mics on a stereo bar in an ORTF-type configuration, slightly below the guitar and facing slightly up. The mics in this configuration will be about 24 to 30 inches away from the player.

Technique #5: As a spot mic in an ensemble, a large-diaphragm condenser 8 to 18 inches above the instrument can be very effective.

GLOCKENSPIEL (ALSO KNOWN AS GLOCK OR ORCHESTRAL BELLS)

The glock is a smaller, higher-pitched version of vibes. Glocks have become quite common in popular music, showing up in almost all genres from hip-hop to jazz.

Considerations

- The glock projects a lot of high-frequency information, so a darker mic, such as a ribbon or dynamic, does a good job of capturing it.
- For less attack, have the player use rubber mallets.

Placement

Technique #1: Place a pair of mics at either end of the instrument, about 6 inches above the bars.

Technique #2: If there are too many transients, place the same mics at the same distance from underneath the glock.

GONG

There are three types of gongs: suspended gongs; bossed gongs, which are flat with a raised center; and bowl gongs, which rest upon cushions. Suspended gongs that are played with a mallet are the ones most used in modern music.

Considerations

A gong has a thunderous low end, so a cardioid mic with proximity effect will make it sound muddy. Try an omni instead.

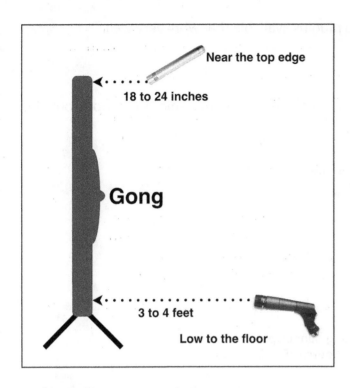

Placement

Technique #1: Place an omni mic about 2 feet from the center at the front of the gong, taking care not to get in the way of the performer. If that placement is a problem, try placing it in the rear of the gong instead.

Technique #2: Place a dynamic mic low to the floor and about 4 feet away to capture the low end, and a small-diaphragm omni condenser near the top of the gong, about 18 inches away, to capture the harmonics (see Figure 9.28).

Figure 9.28: Miking the gong with two mics
© 2023 Bobby Owsinski (Source: iStock Photo)

HAND CLAPS

Claps are best done in a group, and the more clappers the better. Claps are often augmented with foot stomps, boards, or electronic claps to achieve the proper effect.

Considerations

- Use a fairly live room, back the mike away from the clappers, and use a compressor that has variable attack and release times. Compress heavily (10 to 20dB of gain reduction) with a fast to medium attack time so that there's not too much attack. Set the release time fairly fast.

- Gating (but not too tightly) will help keep the track clean. If you need an "ultra-tight to the snare" sound, key the gate from the snare drum.

- Double, triple, or quadruple-tracking the claps makes them sound far bigger.

- As with other percussion instruments, the peaks are always 10 to 15dB greater than what a typical VU meter is reading (if you happen to be using one). Therefore, it's best to record the signal at about -20VU, and -10dB on a DAW peak meter.

- Try having the clappers double the clap track by sitting and slapping their thighs. You get twice as many claps and a somewhat darker, more full-bodied timbre, but not as much edge or definition.

- Space the clappers a couple of feet apart distance-wise from the mic. Five people works well. Have the folks nearest the mic clap more on top of the beat and the folks farther away clap more laid back. Split the claps between two styles: sharp slap-clap done by slapping the fingers of one hand into the palm, and a deeper, more standard palm clap done by clapping both palms together. The distance and timing variation gives a nice thick, cascading effect.

- There are some excellent sample-based clap plugins available that work very well.

Placement

Technique #1: Claps need distance to develop, so start with a dynamic mic (to smooth out the transients) at least 3 feet away (see Figure 9.29).

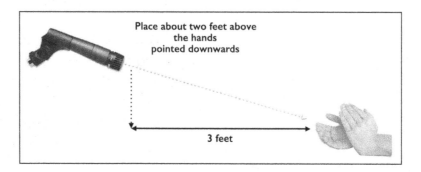

Figure 9.29: Miking handclaps
© 2023 Bobby Owsinski (Source: iStock Photo)

Technique #2: Use a mic with a figure-8 pattern to avoid any slap echo from the floor or ceiling.

Technique #3: Mix in someone clapping two pieces of wood together (instead of their hands) with the real handclaps (a classic old-school trick).

HARMONICA

Miking a harmonica acoustically is different from miking a blues harp, where you're miking an amplifier like you would with an electric guitar.

Considerations

- The harmonica is typically on the shrill side, so a darker mic like a ribbon or dynamic can smooth some of it out.
- Open-ear headphones or using just one phone help the player stay in tune.
- The higher the key, the more shrill the instrument will become.

Placement

Technique #1: Place the mic about a foot away but underneath the hands and pointing up at the harmonica. This will reduce any mouth noises present.

Technique #2: Some players get great results by simply playing into an SM58 cupped in their hands like on stage.

- **Variation:** A combination of both of the above can be very effective.

HARP

The concert harp seen in orchestras has either 46 or 47 strings and seven pedals that are used to change their pitch. The lowest strings are copper or steel wound with nylon, while the middle are gut, and the highest are nylon.

Considerations

- When miking an orchestral harp (not to be confused with a Celtic harp), the main thing to remember is that the sound comes from the soundboard, not from the strings.
- Generally, the major problem in miking a harp is isolation from other instruments.
- For a more natural sound, back up the mic a few feet and let the room support the sound of this very resonant instrument.
- Transient response is less important from a distance (3 feet away or more), but the level drops quickly, so you'll need a quiet room and a low-noise mic and preamp.
- Close-miking requires a mic with good dynamic range to handle the pluck and the ringing (it's like recording a piano, just quieter).

Placement

Technique #1: The classic harp miking calls for a mic placed about 2 feet to one side, about a foot forward of the harpist, and about 4 feet off the floor. This prevents capturing the pedal noise and gets a percussive attack from the fingers.

Technique #2: In orchestral situations, place a figure-8 mic pointed at the middle of the soundboard. Point the null of the mic toward the loudest instruments.

Technique #3: In orchestral situations, clip a small lavaliere mic into one of the sound holes along the musician-side of the instrument. Gain before feedback and ambient noise are never a problem with this setup.

Technique #4: Place a small-diaphragm condenser aimed two-thirds up the soundboard at a distance of about a foot away.

Technique #5: Aim a small-diaphragm condenser slightly above the instrument and to the right, pointed down toward the higher strings.

> **TIP: For overdubs against an orchestra, place a mic where the harp will sit later. Send a mix out of a speaker in the room after the string section leaves but minus the mic sound that was picked up by the harp mic. Reverse the phase of the mic for the harp overdub. The harpist can now hear the orchestra without having to wear headphones and can play more accurately as a result.**

INDIAN INSTRUMENTS

Since most of the readers of this book will rarely be called on to record Indian instruments, we'll group them into a single section. The most common Indian instruments are the sitar, tabla, and tambura.

Considerations

- Remember that the room is half the sound in the case of Indian instruments.
- Sitars produce little sound-pressure level, and the sound tends to emanate from the whole instrument and not a localized area, like a flat-top guitar.
- When recording sitar, remember that the sympathetic strings are heard from far away. Close-miking may destroy this effect, since the close sound is very different from the distant sound.
- The sitar has a very odd radiation pattern, so if you close-mike it, you'll need multiple mics.
- Tabla is a pair of drums. It consists of a small dominant-hand drum called Dayan and a larger metal one called Bayan.
- Many Indian musicians don't seem to think of the tambura as a real instrument, since it's a part of the ensemble for drone ambiance and pitch reference for the vocalists.

Placement

Indian instruments are similar to other wooden string and drum instruments in that they need some distance from the mic for the sound to develop.

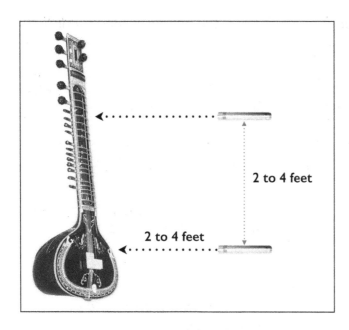

SITAR

Technique #1: For sitar, use a pair of omnis at a distance of 4 to 10 feet. Mic positioning can generally be from about a foot to about 3 feet off the floor with the mics pointing down at the instrument (see Figure 9.30). Greater heights can produce a more unpredictable effect.

Figure 9.30: Typical sitar miking
© 2023 Bobby Owsinski (Source: iStock Photo)

Technique #2: For close-miking a sitar, place a dynamic mic in close to the bowl and a small-diaphragm condenser aimed at the neck to get a thicker sound.

Technique #3: For either a pure solo or a traditional ensemble context, an X/Y stereo pair of microphones placed 3 to 4 feet in front and slightly above the performer, angled toward the instrument, can produce a larger-than-life sound.

Technique #4: Place an omni behind the performer, almost looking over the player's shoulders, and a cardioid approximately 2 feet in front and angled back toward the picking hand.

TABLA

Technique #1: Place an dynamic mic on a small stand a foot away and slightly above the tabla, between both drums. You may have to move it slightly toward the smaller drum to ensure that the recording level of both drums are the same.

TIP: Sometimes it's best to use a windscreen on the microphone to protect it against the talcum powder that tabla players use to dry their hands.

- **Variation 1:** Instead of pointing the mic at the drums, point it downward at the floor, halfway between the drums and the mic in order to reduce the attack.
- **Variation 2:** Change to an omnidirectional mic instead.

Technique #2: Place two mics in an X/Y stereo configuration 2 feet back and 2 feet up from the tabla.

- **Variation:** Add additional spot mics on each drum.

TAMBURA

Technique #1: For tambura, place a small-diaphragm condenser mic about 18 inches away from the body. You don't want to hear the individual plucking of the strings, just the resulting drone.

KOTO

The koto, the national instrument of Japan, is traditionally a 13 string instrument, but some may have as many as 25. Although it's thought of as a distinctly Japanese sound, it has been used in popular Western music by David Bowie, Queen, Dr. Dre, and the jazz-rock band Hiroshima. The player uses three finger picks (thumb, forefinger, and middle finger) to pluck the strings.

Considerations

- The koto has a lot of high-frequency energy, so a darker mic like a ribbon or dynamic captures it well.
- The tuning is adjusted by moving the bridges before playing.

Placement

Technique #1: Place a single small-diaphragm condenser a foot above the player's head, directly over the instrument, aiming straight down.

- **Variation:** Instead of a single mic, use two in an ORTF or X/Y stereo configuration.

Technique #2: Place a lav or contact mic inside the koto and a mic on the koto's sound hole. This works especially well in a live situation.

LESLIE SPEAKER

The signature sound of the Hammond organ (and many other organs, for that matter), is the Leslie speaker, which is a wooden cabinet with a set of rotating speakers that employ acoustic Doppler shift to obtain their effect. The high-frequency horn provides frequency modulation (Doppler shift), while the rotating low-frequency section provides amplitude modulation.

Considerations

- The farther away the mics are placed from the speakers, the less you'll hear the noise of the Leslie motors.
- The Leslie rotating effect is much more dramatic when the louvers are miked, rather than the horn opening.

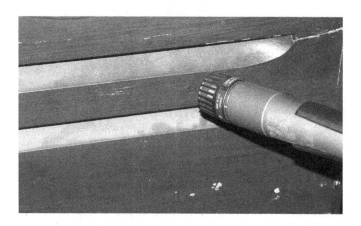

Placement

Technique #1: Place a mic directly on the top louver of the Leslie and another on the bottom louver (see Figures 9.31 and 9.32). Placing it on the louvers instead of the cabinet openings intensifies the effect and decreases the wind noise.

Figure 9.31: SM57 placed at the top Leslie louver
© 2023 Bobby Owsinski

Figure 9.32: AKG D 112 placed on the bottom Leslie rotor
© 2023 Bobby Owsinski

- **Variation 1:** The classic stereo method: Place two mics at opposite sides of the cabinet against the top louvers and put a single large-diaphragm condenser on the bottom louver. Since the two mics on top are on either side of the cabinet, you hear the moving Doppler effect when they're panned to different places in the stereo soundfield.
- **Variation 2:** As above, but place one of the upper mics on the front louver. Be careful if this will go to mono, since the effect of the top mic will diminish because of the phase cancellation (see Figure 9.33).

The Recording Engineer's Handbook - 5th edition

Figure 9.33: Leslie top and bottom rotors miked in stereo
© 2023 Bobby Owsinski

- **Variation 3:** Add a distant mic approximately 4 to 6 feet away, aimed at the top rotor. The bottom mic is panned center, the top close mic is panned to 11:00, and the distant mic is panned to 3:00, mixed at least 6dB lower in level than the close mic.
- **Variation 4:** Place a pair of small-diaphragm condensers 8 to 10 inches away from each side of the Leslie, at about the middle and pointed slightly up.

Technique #2: Place a mic about 5 feet away from the side of the Leslie, aimed about halfway down the cabinet.

Technique #3: Place an omni directly on the top of the Leslie cabinet, using small, folded pieces of a matchbook or a masking-tape reel to isolate the mic from vibration. This requires a moderately reflective room.

MANDOLIN

Mandolin is basically a soprano lute. It has four courses (double strings) that are tuned the same as a violin and opposite of a bass (G D A E).

Considerations

- Watch the hand movements of the player so the picking hand doesn't obscure the mic.

Placement

Technique #1: Use one mic pointed down at the top string from about 6 inches away, and a second mic pointed up at the sound hole from underneath, again at a distance of about 6 inches.

Technique #2: Place a mic between where the picking hand and the neck meet, at a distance of between 12 and 18 inches. Place a second mic near the lower f-hole for fullness.

Technique #3: For stereo, place a mic on each f-hole, keeping them about 6 inches apart.

MARIMBA (ALSO SEE VIBES)

A marimba is a type of xylophone, but with broader and lower tonal range and resonators. It consists of a set of wooden bars with resonators that are arranged like the keys on a piano.

Considerations

- A good marimbist can produce a very large dynamic range.
- The tonal character can change completely depending upon the type of mallets used. A marimbist has to use soft sticks at the bottom end of the marimba, and if the music calls for him or her to shift to a much higher register without the time to change sticks, it's likely that you will pick up a considerable amount of noise from the instrument's frame.
- The sound of a marimba doesn't come from a single location; it builds up throughout the instrument and is usually largely dependent on the natural acoustics of the room.
- Marimbas have a lot of individual parts, which often results in various audible buzzes or rattles from the frame, the resonators, or the bars and support posts.

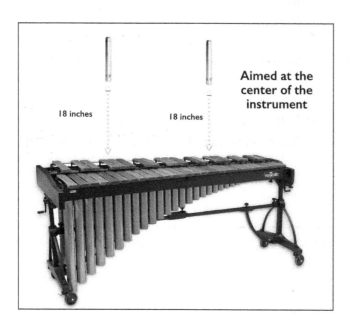

Placement

Technique #1: Place two small-diaphragm condensers about 18 inches from the keys at about a 45-degree angle (see Figure 9.34).

Figure 9.34: Typical marimba miking
© 2023 Bobby Owsinski (Source: iStock Photo)

Technique #2: To record in mono, place a mic 3 feet over the center of the instrument.

MOUTH HARP (ALSO CALLED JAW HARP)

Said to be one of the oldest instruments in the world, the mouth harp has a metal or bamboo tongue reed attached to a frame.

Considerations

- It's possible to modify the pitch of a mouth harp by varying the shape of the mouth.
- The instrument is very quiet and requires much more amplification than other instruments, so the level of noise of both the mic and the preamp becomes a factor.

Placement

Technique #1: Place a mic (a dynamic works fine) from 12 to 18 inches from the nose of the player, pointed at the harp.

OBOE (ALSO FOR ENGLISH HORN)

Many consider the oboe to be one of the most difficult instruments to learn how to play, with a huge learning curve requiring an expensive instrument that can take up to 5 years to make. Unlike other musical instruments, oboes sometimes have their wood crack during use. African hardwood works best but it's only available from one forest in Tanzania, and it's endangered so only trees that are 60 years old or older are harvested.

Considerations

- Most of the sound comes from the body of the instrument, not the bell.
- The keys can make clicking noises when the instrument is played, which can be picked up if close-miked.
- Although the frequency response from an oboe can extend out to 14kHz, there's virtually nothing below 220Hz
- The oboe is frequently used to set the tuning of an orchestra because its tone is so stable and its piercing sound can be easily heard.
- The tuning changes with temperature, and it goes sharp as the air gets warmer.

Placement

Technique #1: Place an omnidirectional microphone about 2 feet away pointed about 1/3rd of the way up from the bell.

Technique #2: For more isolation and less room ambiance, place a small directional clip-on microphone to the bell of the oboe and point it down the tube at the keys. Experiment with mic position to find the sweet spot.

Technique #3: If there is more than one player, place the microphone in between two players over the oboes and angled a bit downward.

PENNY WHISTLE (TIN WHISTLE)

Closely associated with Irish traditional music and Celtic music, the penny whistle is a simple six-holed wind instrument that's in the same class as the recorder. The most common whistles today are made of brass or nickel-plated brass with a plastic mouthpiece.

Considerations

- The penny whistle has a lower output in the lower registers, so it helps if the whistle player leans in towards the microphone during those passages if the mic is close.
- A dynamic or ribbon mic is preferred so the shrill nature of the instrument isn't magnified.

- Pointing the mic at the sound hole will capture too much breath noise.
- Attenuate the 2k to 3kHz region if the recorded sound is too shrill.

Placement

Technique #1: Place the mic at least 2 feet from the instrument pointed toward its center.

PIANO (GRAND)

The piano is a relatively new instrument, dating back to only around 1700. Although it comes in the most familiar grand and upright versions, there are subcategories to each.

A concert grand is generally between 7 and 10 feet long, a parlor grand is 6 to 7 feet, and the baby grand is about 5 feet long. Upright pianos come in a studio version (42 to 45 inches tall), the more compact console version (38 to 42 inches), and the spinet version, where the top barely rises above the keyboard. All else being equal, longer pianos have longer strings and richer tone.

Considerations

- When microphones are placed inside the piano, they can pick up unwanted pedal and hammer sounds in addition to the music, but they'll also capture a brighter, closer sound.
- Microphones placed outside but near the side of the instrument "looking in" can also capture reflections from the piano's top. That can be good or bad depending on the sound you're looking for.
- Microphones placed away from the instrument will record both the piano and the room. If your room sounds good and you don't need a very close sound, this is a safe method for recording a balanced piano, as the sound of the instrument doesn't really exist properly until you get some distance from it.
- Miking from the side usually means that the higher notes will be louder. Miking inside the case will tend to emphasize the middle octaves, which could be good for some music styles and not for others.
- To make the piano brighter, add a few dB at 10kHz. For more definition, add a little at 3kHz. If the piano sounds thin, adding a few dB at 100Hz helps.
- Make sure to hire a professional piano tuner before your recording session to be sure that the piano will be in tune.

Placement

Technique #1: For orchestral or solo piano, have the pianist play scales. Stick your finger in one ear and walk around to find the point in the room where the hall ambiance and the direct piano sound are balanced. When you find that place, walk back and forth along the piano and listen to how the low end changes. Find a place that has the right tonality and the right ambient balance. Place an ORTF pair there.

Technique #2: Place a spaced pair of mics about 6 feet away with the mic closest to the player pointing at the hammers and the low mic pointing at the low strings.

- **Variation 1:** Place a pair of small-diaphragm condensers spaced about 4 feet apart and positioned about 6 feet away from the piano and about 8 to 9 feet high. Angle them to aim at the edge of the lid when on full stick.

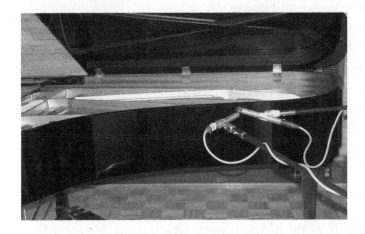

Technique #3: Place an X/Y pair aimed at the middle of the "rounded" part of the piano at an angle equidistant between the keys and the lid (see Figure 9.35). If there's not enough ambiance from this technique, move the pair back and up but keep the mics aimed at the same place.

Figure 9.35: Small-diaphragm microphones in an X/Y configuration placed outside of the piano
© 2023 Bobby Owsinski

- **Variation 1:** With the piano lid at full stick, place a pair of X/Y cardioids at the edge of the body, aimed in and down more toward the higher strings. (The lower end of the piano is omnidirectional and will be picked up as well.) Move as needed to balance the sound.

- **Variation 2:** Place a pair of figure-8 mics in a Blumlein configuration 8 or 10 feet outside the piano, high enough to be pointed toward where the hammer hits the string at about C above middle C.

Technique #4: For more of a rock sound, place a mic about two-thirds of the way up the bass strings, 12 to 18 inches over the strings, and another aimed at the label (under the high strings) from a similar height. To balance between the high and low strings, put the monitors in mono and move the mics around until it sounds balanced, then record on two tracks.

- **Variation 1:** Place one mic by the upper mid hammers and one by the extreme back end over the lower strings (see Figure 9.36).

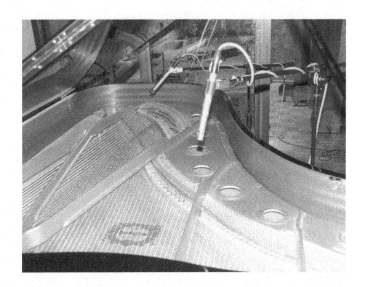

Figure 9.36: Small-diaphragm condenser microphones inside the piano.
© 2023 Bobby Owsinski

187 | Miking Individual Instruments

- **Variation 2:** Also a classic rock-and-roll setup, place an X/Y or a near-coincident pair within 2 feet of the center of the harp, where the high and low strings cross (see Figure 9.37).

Figure 9.37: Small-diaphragm condenser microphones in an X/Y array, placed inside the piano
© 2023 Bobby Owsinski

- **Variation 3:** For more body, add another mic in one of the classic positions outside the piano in the room. Again, listen in mono for correct balance and phase.

Technique #5: Sometimes recording from above and behind the pianist works well—especially with a stereo pair. Point the mic toward the soundboard, about an octave below middle C. You don't get much attack from the strings, but it captures a good fill sound that sits well in the mix.

- **Variation:** Place the mic just above the music stand, aimed directly in the center of the soundboard.

PIANO (UPRIGHT)

There are numerous ways to mike an upright piano, none of them particularly ideal. The way that sounds the best (Technique #2 below), frequently suffers from stool and foot noise from the player.

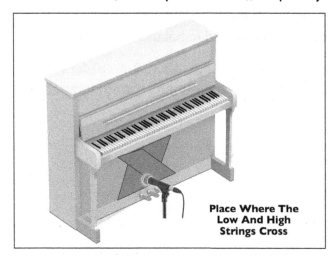

Placement

Technique #1: Place an ORTF or X/Y stereo pair about 3 feet away from the back soundboard.

Technique #2: Take the panel off underneath the keyboard. Place a mic (omni works well) 12 inches away from the point where the low and high strings cross (see Figure 9.38).

Figure 9.38: Miking an upright piano underneath the keyboard, where the strings cross
© 2023 Bobby Owsinski (Source: iStock Photo)

- **Variation:** Add another mic right above the separation, pointed toward the high keys.

Technique #3: Open up the top of the piano and place a mic on the bass side and another on the treble side facing down, just over the lip of the piano.

PICCOLO

The smallest of the flutes, the piccolo can be constructed of wood, metal, gold, plastic, or any combination.

Considerations

- Because the piccolo's sound is in such a high register, it has the tendency to be shrill, so a darker-sounding mic can be very effective in capturing its true sound.

Placement

Technique #1: Place the mic about 10 feet away and pointed at the middle of the instrument. It also helps to compress the signal 2 to 4dB at about a 4:1 ratio to smooth out the variations in level between notes.

RECORDER

The recorder is a whistle-like woodwind instrument that was the precursor to the flute. Its use has been revived in the 20th century to keep performances historically accurate and as a simple instrument for teaching music. It's also been used in modern music on recordings by David Bowie, The Beatles, Led Zeppelin, Kate Bush, Bruce Springsteen, and the Rolling Stones, among others. Most high-quality recorders are made from a range of hardwoods, such as maple, pear wood, rosewood, grenadilla, or boxwood with a block of red cedar wood, although inexpensive plastic recorders are used as well.

Considerations

- The sound of the recorder is very pure and has a distinct lack of upper harmonics.
- Recorders come in a variety of sizes, and each has its own register, although most are tuned in C or F.

Placement

Technique #1: Similar to a flute, place a mic from 1 to 2 feet over the middle of the instrument, looking down.

SAXOPHONE

The sax is another relatively new instrument, invented in the 1840s as a means for woodwinds to keep up with the brass instruments in volume and projection.

Considerations

- The sound of a saxophone comes from every hole and the body of the instrument at the same time, but in totally different proportions for every note. The bell gives you the honk on the highs and some of the low-frequency components, with only the lowest note coming exclusively out of the bell.

- The notes at the top of the instrument range come out of the upper body, left-hand side. Altissimo notes (higher again), typically high-pitched screams, come out of the front upper and middle tone holes but are usually much louder than most other notes.

- The bell sound is generally quite focused but disproportionately edgy and harsh. The side pads of the saxes generally radiate a "woody" sort of tone, which by itself can sound like the reed is soggy.

- When they're warming up, sax players will inevitably find the spot in the room (usually a couple of feet from a wall) where the horn sounds best to them. This may be the best place to put the player or the mic.

Placement

Technique #1: Place a mic directly in front of the sax at a distance of 12 to 16 inches to capture a sound that's very authentic.

- **Variation 1:** Place a second mic up at the top of the sax, up on the left of the instrument, near the reed. Since the sound coming from this placement might be a bit harsh, try using a mellower-sounding mic, such as a ribbon or a dynamic. Balance the sound between both mics to get the ideal blend. Move the mics back slightly for more room ambiance and to decrease the valve clicks.

Technique #2: Place a mic about 18 to 24 inches away from the player's left side of the instrument, about halfway up the keys, and aimed slightly down at the bell (see Figure 9.39). Have the sax player play the song that you'll be recording. Try moving the mic 6 inches closer to the sax for a tighter sound, but be careful not to pick up the valve clicking as well. If you move the mic away from the sax, you'll pick up more of the room ambiance.

Figure 9.39: Miking a sax
© 2023 Bobby Owsinski (Source: iStock Photo)

Technique #3: Place a mic at about neck or face high of the player, aiming down at the horn.

Technique #4: Position the mic about 3 feet away, slightly above head height and about 30 degrees to left of the player, aimed toward the middle of the horn.

Technique #5: For soprano sax, place the mic above the sax at about the midpoint, aimed straight down.

SHAKER

Shakers have become a vital part of the rhythm section in popular music, subtly adding motion to a song. Many drummers and percussionists use a wide variety of shakers, both store-bought and homemade.

Considerations

- There are many different types of shakers, and they all sound different. However, the technique for capturing their sound remains the same.
- Sometimes some salt inside two Styrofoam drinking cups taped together will sit better in the track than a real shaker.
- As in miking a tambourine, the mic placement has to be far enough away so you don't pick up any of the movement or wind from the player.
- Most shakers have high-frequency transients that place extra strain on the microphone preamp. Try several (if you have them) to find the combination of mic and preamp that sounds the best.

Placement

Technique #1: Set a condenser mic to omni and place it 5 or 6 feet away from the shaker at about head level (see Figure 9.40).

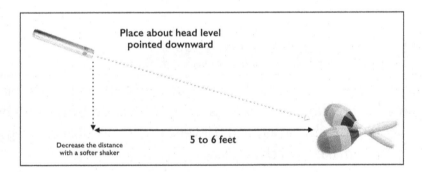

Figure 9.40: Miking a shaker
© 2023 Bobby Owsinski (Source: iStock Photo)

- **Variation:** As above, only using a ribbon mic instead.

Technique #2: Place a mic behind the player and from 1 to 2 feet over his head, aimed down at the shaker.

STEEL DRUMS (ALSO KNOWN AS STEEL PANS OR PAN DRUMS)

Originally made from a 55-gallon oil drum but now machined to precise specifications, steel pan drums are chromatic and have 13 members in the pan family. The most commonly used is the double tenor.

Considerations

- Steel drums are very much like fiddles in that their radiation pattern changes with every note played. There is no single place in the near field where you can mike it and have all the notes captured evenly.

- A close-miked steel drum can sound more clanky than melodic.

- Have the player rewrap his pan sticks (or replace the surgical tubing on the end, if they're constructed that way) for the recording. Having fresh mallets to work with makes a huge difference.

- Dynamic mics often work better than condensers because they smooth out transients from the drum's attack.

Placement

Technique #1: Place a ribbon mic directly over the player's shoulder.

Technique #2: Place an omni about 6 feet away from the pans and about 2 feet above them.

Technique #3: Place an omni under the rim of the steel pan, pointing up at the center.

Technique #4: Place a dynamic mic about 8 feet above the pans to allow the transients some space to dissipate.

STICK (CHAPMAN STICK)

The Stick (sometimes known as the *Chapman Stick* after its inventor, Emmett Chapman) is a 10-stringed instrument, although versions with 8 or 12 strings also exist, played by tapping both hands on the fretboard. It gives a player the ability to play chord inversions not possible on a regular guitar, and for an instrumentalist to play both bass and chords/lead at the same time.

Considerations

- The Stick is an electric instrument with two outputs: one for the high strings and one for low strings.

- The strings on the bass side vary in volume as you move up the neck, so a bit of compression is useful to match the level of all the notes.

Placement

Technique #1: Use two DIs for the outputs.

Technique #2: Connect the high-string output to a guitar amp and the low-string output to a bass amp (or DI) and proceed as if miking a normal guitar and bass amp.

STRING SECTION

Since the majority of people who read this book won't ever get the chance to record a large symphony orchestra (and probably aren't interested in doing so either), we won't go into that much in this book. There's plenty of information available in the many books dedicated to this subject. Plus, you can read the interviews with Bruce Botnick, Michael Bishop, and Eddie Kramer, which outline the major approaches. However, recording a small string section is within the realm of possibility for most engineers, so here are some techniques.

Considerations

- Before recording, go out in the room and listen to the sound of the string section next to the conductor (if there is one) to hear what he or she hears.
- All string instruments radiate omnidirectionally, but the brilliance of the tone comes from the top of the instrument. Violins and viola (known as the chin strings), project up and over the performer's right shoulder. The celli and basses project more forward and lower.
- You can record a bowed string instrument from any angle, but the results are usually better if the mic can "see" the top of the instrument.
- When recording an ensemble, walk around it to find where the group sounds best, then put the mics in that sweet spot.
- Go with closer miking (but not too close) if the room is small, since the distant approach may show how small the room is and pick up unwanted reflections.
- Spot mics are okay to use since it's always easier to not use a mic during the mix then to not have the control you need.
- The lush sound that the composer expects comes from the front mics placed above the conductor, but the clarity of the inner voices comes from the spot mics.
- A good room is a necessary ingredient for a good string sound. A low ceiling with acoustic tile will murder your string sound (and every other sound, too).
- Strings do not sound beautiful when close-miked. The sound is usually harsh and shrill. Strings need space for the sound to develop.

Placement

Technique #1: When there is an even number of violins, violas and cellos (like the typical 4-2-2 or 8-2-2 arrangement), place a mic 3 or 4 feet above and in between each pair of violins and violas, then a separate mic 2 to 3 feet away from each cello.

Technique #2: Place an X/Y pair a few feet behind the conductor (who stands about 6 feet away) and about 10 to 12 feet high, but not too close to the ceiling (see Figure 9.41).

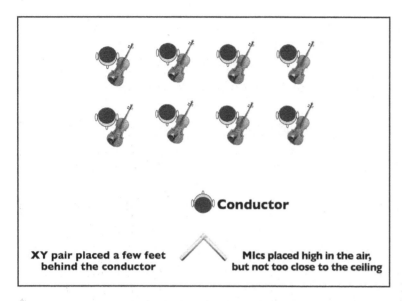

Figure 9.41: Miking the string section with an X/Y pair
© 2023 Bobby Owsinski (Source: iStock Photo)

- **Variation 1:** Add individual "sweetener" mics for each section to the X/Y pair. Get the overall sound from the stereo mics and then add the sweeteners when mixing.

- **Variation 2:** Other stereo configurations, such as ORTF, Blumlein, or a Jecklin Disk, can also work well in this position.

Technique #3: Use a spaced pair about 20 feet apart and at least head height or higher, pointing down at about 30 degrees.

Technique #4: For string quartets, place a small-diaphragm condenser on each player about 3 feet away, but more out front than overhead. Supplement these with an ORTF pair of large-diaphragm condensers about 7 feet high and about 3 or 4 feet back of the line between the first violin and cello. Move the mics closer or farther away to balance the amount of room and direct sound.

Technique #5: Most large orchestra recordings still uses a Decca Tree that sets three omnidirectional mics over the conductor's head in a T arrangement (see Chapter 6). The mics used for a Decca Tree are usually Neumann M 50's because of their increased high-frequency response in the diffuse field of a hall.

SYNTHESIZER (OR ANY KIND OF ELECTRIC KEYBOARD)

While it's fairly easy to take a synthesizer or electric keyboard direct, many times you can enhance the sound by adding some natural room sound along with the direct sound. Sometimes this is the difference between a synth sounding real or artificial.

Considerations

- The reason why electronic keyboards don't sound "real" is that they don't have a first reflection of the ambient soundfield that all acoustic instruments have.

- Many so-called "stereo" instruments achieve their stereo spread by modulating one of the outputs. This can cause phase problems when summed into mono. Sometimes you're better off using only the output labeled "mono" and achieving the stereo spread another way during mixing.

Placement

Technique #1: The usual method of recording a synthesizer is to take the outputs directly into the console or DAW through direct boxes.

Technique #2: Another way is to feed the keyboard signal back out into the studio (by either the playback speakers or a guitar amplifier) and mike the room from a distance of about 6 to 10 feet with a stereo mic, X/Y, or ORTF pair (see Figure 9.42). This accomplishes two things: It adds the needed first reflections to make the instrument sound more real, and it provides a nice stereo spread that sounds a lot more natural than the artificial stereo found on most synths and sums to mono well. Add this to the direct sound.

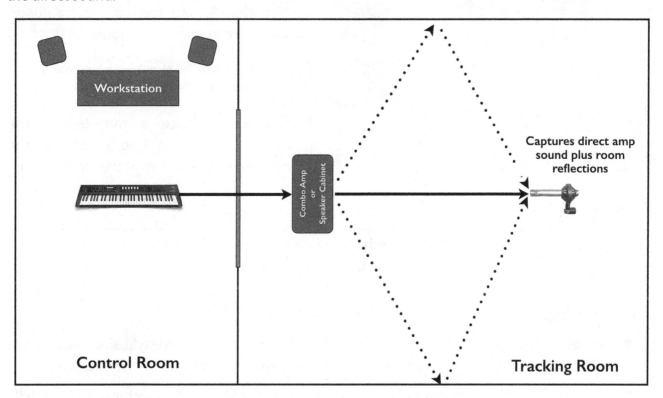

Figure 9.42: Feeding the synth into the room
© 2023 Bobby Owsinski (Source: iStock Photo)

TAMBOURINE

Tambourines come in many sizes and shapes, and are made out of a variety of materials. The classic instrument has a drumhead, but most modern varieties used in pop music do not.

Considerations

- Place your tambourine player in the livest, brightest part of your recording room.
- Keep in mind that since a tambourine must be moved when playing, close-miking might pick up a lot of air noise or movement from the percussionist.

- If you're using an analog console that uses VU meters, try to keep the level to around −20dB VU, as a tambourine has peaks that can be a full 15dB over what a VU meter is reading.
- Sometimes a really dark-sounding mic (something with a rolled-off high-frequency response) works best.

Placement

Technique #1: Place an omni mic 5 or 6 feet away at about head level.

- **Variation:** Replace the omni with a ribbon mic and place as above (see Figure 9.43).

Figure 9.43: A Royer R-121 miking a tambourine.
© 2023 Bobby Owsinski

Technique #2: Place a mic behind the percussionist and from 1 to 2 feet over the player's head, aimed down at the tambourine.

TIMBALE

Typically, a timbale set consists of two single-headed drums, a wood block, and two cowbells mounted between the drums. A small cymbal may also be mounted off to the side.

Considerations

- Timbale miking can change considerably if the timbale is used in Latin style, since the cowbell will be attached and can easily overwhelm any mics aimed at the top of the instrument.
- If the music is Salsa, the player may play on the side of the drum, which can only be fully captured from below.

Placement

Technique #1: Place a dynamic mic (an SM 57 works nicely) under each drum, about 6 to 12 inches away, angled at the head.

- **Variation 1:** To capture the entire timbale kit, add a small-diaphragm condenser mic to the under mics placed over the set at a distance of 1 to 2 feet. Position the mic for best balance between wood block, cowbells, and cymbal.
- **Variation 2:** Replace the dynamic with a ribbon mic. The figure-8 pattern will capture more room sound, which may help with the overall sound if the room sounds good.

Technique #2: Place an ORTF or X/Y pair about 2 feet over the timbale kit to capture a nice stereo picture. Move the mics as needed to get the best balance.

Technique #3: For just a mono recording of the kit, place a mic from 1 to 2 feet away, as above.

TIMPANI (ALSO KNOWN AS KETTLE DRUM)

Although primarily an orchestral instrument, timpani has been used in modern music by Elvin Jones, The Beatles, Led Zeppelin, Queen, and the Beach Boys, among others.

Considerations

- Unlike most drums, timpani produce a definite pitch when struck.
- The shape of the bowl determines the sound of the drum. Parabolic bowls are usually brighter than hemispheric bowls.

Placement

Technique #1: Place a large-diaphragm condenser from 3 to 6 feet directly over the drum, looking down.

Technique #2: Place a dynamic mic (an RE20 works great) about a foot away and a foot underneath the drum. Point the mic up so that it just peeks over the head. Add a large-diaphragm condenser directly over the drum looking down, as in Technique #1.

TRIANGLE

A triangle is usually made of steel and left open at one end so that it doesn't have a true pitch.

Considerations

- The triangle has a lot of transients, a lot of harmonics, and a lot of high-frequency energy, which makes it one of the hardest instruments to record. You can really tell the differences between preamps as a result (some engineers use a triangle as a test before buying). Try several different preamps to see which sounds best if you have them available.
- Microphones can be really stressed to their limits when recording a triangle. A dynamic mic sometimes works well, since it softens the transients.
- Make sure to engage the high-pass filter on either the mic, the preamp, or the console in order to decrease the low harmonics and any unwanted low-frequency noise or rumble.

Placement

Technique #1: Place a small-diaphragm condenser mic in *omni* at least 3 to 4 feet away, slightly above the triangle and aiming down. Move the mic back until any low-frequency harmonics or microphone distortion disappears.

Technique #2: Place a mic behind the player and from 1 to 2 feet over his head, aimed down at the triangle.

TROMBONE

See the "Brass" section.

TRUMPET

See the "Brass" section.

TUBA

See the "Brass" section.

UKULELE

A relatively new instrument, the ukulele originated in Portugal and was introduced soon after to the Hawaiian Islands, where it gained popularity. Although the instrument is made of many materials, the preferred wood is koa, which is endemic to Hawaii.

Considerations

- The nylon strings of the ukulele don't produce a great amount of overtones, which makes the sound warm and mellow.
- Ukuleles come in different sizes and pitches. The soprano ukulele is the most common.

Placement

Technique #1: Place a directional microphone about a foot away from the 9th fret and aimed towards where the neck meets the body.

VIBES (OR VIBRAPHONE)

The vibraphone is similar to the xylophone, marimba, and glockenspiel, except that each bar is paired with a resonator tube that has a motor-driven butterfly valve at its upper end. This produces a tremolo effect (a periodic change in volume) while spinning. The vibraphone also has a sustain pedal similar to that used on a piano.

Considerations

- Experiment with the position of the fans. Setting them vertically will increase the volume of the resonators and decrease the sustain of the instrument; setting the rotors flat will decrease the volume of the resonators but increase the sustain.

- If the player isn't using the motor for the tremolo effect, make sure that the rotor fans in the resonators are set to the same angle for each set of resonators, for consistency across the range of the instrument and between the upper and lower manuals.

- Be prepared to spend a few minutes getting rid of squeaks and rattles.

Placement

Technique #1: Place a spaced mic pair or an ORTF pair about 2 to 3 feet over the center of the instrument, dividing it equally.

- **Variation:** For a closer sound, add an omni about 2 feet off the low F bar in the bass end.

Technique #2: Place a mic over each corner and one over the center, about 6 to 8 inches above the bars. Keep them just out of mallet reach, tilted slightly downward and aimed into the nearest third of the instrument. Place the center mic at the same height, but back just a little compared to the other two and aimed at a slightly greater angle down and toward the middle of the instrument.

Technique #3: Place four mics with the two splitting the middle and the outer two covering a bit less of the instrument.

VOCALS (BACKGROUND)

Recording background vocals is distinctly different from recording solo vocals because of how they fit in the mix.

Considerations

- Since background vocals are invariably stacked, layered, or at the very least doubled, try the following to make them bigger and have a greater sense of space. For every subsequent overdub after the first recording, have the singers take a step backward, but increase the mic preamp gain so that the channel's level is equal to the first layer. In essence, you want to "fill up the meters" so the level on the meter is the same regardless of where the singers stand.

- If the singers have trouble blending or singing in tune, ask them to remove one side of the headphones or at least put it slightly back on the ear. Sometimes this helps them sing in tune, since they can then hear the blend acoustically.

- The best background blends usually come from having multiple singers positioned around a single mic or stereo pair.

- Large-diaphragm cardioid condensers are usually used because they combine a proximity effect and a slight midrange scoop along with a slight lift in the upper frequency ranges. This accentuates the "air" portion of the sound, which helps the background vocals sit better in the mix.

- The microphone does not always have to be a large-diaphragm condenser. Sometimes the natural compression of both volume and transients offered by a dynamic mic will keep the vocal much more under control than a condenser would.

- The better the lead vocalist is, usually the harder it is for the singer to do background vocals with other people because the voice is recognizable.

- If the lead singer is singing the background parts or is part of the background vocal ensemble, try not to use the same mic that the lead vocals were recorded. This will avoid any buildup of any frequency peaks in the singer's voice, the mic, or the room.

- Always try to do something a little different on each background vocal track. A different mic, mic preamp, room, singer, or distance from the mic will all help to make the sound bigger.

- A trick to help things blend better is to record the background vocals and then play them back through the studio monitors and record the playback. The distance of the mic placement depends upon the sound of the room. Walk around and find the place where the playback sounds the best, and then place the mic there. Record and mix this in low underneath the original background vocal tracks.

- For a performance where the singers will sing a three-part harmony, use a cardioid pattern with three people on the mic. On the first pass, one person is on-axis and the other two are off-axis by 90 degrees (facing each other). On the next pass, have one of the off-axis singers trade places with the on-axis singer. Do a third track the same way until all three singers have been recorded on an on-axis track and all three notes are on-axis. If it's only a two-note harmony, then double each note using the same method.

- Another variation of the above technique is to have all three people sing the same note at the same time using the same mic technique; then proceed as above.

Placement

Technique #1: Place an omni mic from 2 to 3 feet away from the vocalists (see Figure 9.44).

Figure 9.44: Miking the background vocals with a single mic
© 2023 Bobby Owsinski (Source: iStock Photo)

Technique #2: The standard jingle production technique: Stack the overdubs with three vocalists on a pair of mics in X/Y or a stereo mic. Have them sing each part in unison and then change position and double. Do the same for all parts.

Technique #3: For extra-thick background vocals, cut four tracks of the root, two tracks of the harmony, then one or two "whisper" tracks of each part. Compress the whisper tracks heavily.

Technique #4: Many of the techniques used for recording a choir also work for background vocals, such as the Blumlein pair.

VOCALS (SOLO)

There are as many vocal microphones as there are people singing, yet the sound totally depends upon the singer and the delivery. If the singer has bad technique, no expensive microphone or signal path can save the performance. Just like with the various instruments, it really comes down to the talent of the singer more than any one thing when recording a vocal.

Considerations

- A singer who is experienced at working the microphone reacts to what he or she hears in the headphones and balances the consonants against the vowels.

- Decoupling the stand from the floor will help prevent unwanted rumbles. Place the stand on a couple of mouse pads or a rug for a cheap but efficient solution.

- A major part of the silky-smooth hit female vocal sound is singing softly and breathily, very close to the mic. To get this effect, tell the singer to close her eyes and act as if the mic is a baby's ear. This usually produces vocals that are very soft and natural.

- The best mic in the house doesn't necessarily capture the best vocal sounds. Match the mic to the character of the singer's voice.

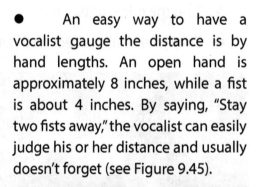

- An easy way to have a vocalist gauge the distance is by hand lengths. An open hand is approximately 8 inches, while a fist is about 4 inches. By saying, "Stay two fists away," the vocalist can easily judge his or her distance and usually doesn't forget (see Figure 9.45).

Figure 9.45: Setting the vocal distance by hand
© 2023 Bobby Owsinski (Source: iStock Photo)

In general, vocals sound better when recorded in a tighter space. Vocal booths should be tight but not acoustically dead to the point where there is a loss of high frequencies (especially the air band). Low-ceiling rooms can also be a problem with loud singers, as they may cause the room to ring at certain lower midrange frequencies.

Figure 9.46: Vocal mic placement using a large-diaphragm condenser mic
© 2023 Bobby Owsinski

201 | Miking Individual Instruments

To Eliminate Pops, Lip Smacks, and Breath Blasts

1. Place the mic above the lips so the singer's breath is right below the capsule (see Figure 9.46).

2. Move the mic up 3 or 4 inches above the singer's mouth and point it down at the lips. This also cleans up mouth noises and the nasal sound that some singers have (see Figure 9.47).

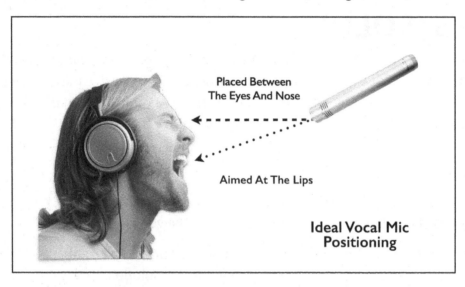

Figure 9.47: Alternate vocal mic placement using a condenser mic
© 2023 Bobby Owsinski

3. If popping continues:

- Move the mic higher and/or farther away.
- Turn the mic slightly off-axis.
- Change the mic's directional pattern to omni.

THE HANGING MICROPHONE

Everyone has seen the photos of the vintage large-diaphragm tube mic hanging upside down in front of the vocalist (see Figure 9.48), but there really are several good reasons for this. Here are just a few things to consider:

Figure 9.48: The hanging tube condenser microphone.
© 2023 Bobby Owsinski

- The rationale behind hanging a mic upside down comes from vintage tube mics. The heat rising from the tube and the internal electronics can cause the diaphragm to change temperature over time, which will change the sound of the mic. Placing the mic so the tube is above the capsule will let the heat rise without passing over the diaphragm.
- Another thing that happens is that the vocalist sings slightly upward into the mic, which forces the airway open and encourages a full-body voice. Take a deep breath and sing a low note, start with your chin to your chest, and slowly lift your head until your chin has about a 15-degree lift. Hear any difference?
- Maybe even more important, the mic can be positioned so the singer is less likely to direct popping air blasts into the mic.
- It's also easier for the singer to read any music or lyrics since the mic is out of the way.
- Modern condenser microphones, even ones with tubes in them, do not suffer from these problems and therefore can be placed in whatever position that's required to capture the best sound.

Placement

Technique #1: Place the mic with the capsule just about even with the singer's nose and point it down at the lips. If popping continues, turn the mic slightly off-axis. The distance will vary widely depending on the singer, the sound you're trying to get, and the SPL handling capability of the mic. Somewhere between 4 and 12 inches should work for most situations. To record a whisper, place the mic even closer than 4 inches to the singer.

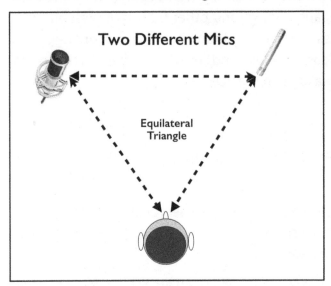

Figure 9.49: Vocal mics in a equilateral triangle
© 2023 Bobby Owsinski (Source: iStock Photo)

Technique #2: If the vocalist has trouble staying in the right place or wants to eat the mic, use a close mic as a decoy and put the one that you're recording with just behind it. While the sound might be a little more distant, it will also be a lot more consistent.

Technique #3: To fill out a thin-sounding voice, position a mic 4 to 6 inches below the vocalist's mouth and then aim the mic up at the lips. You'll pick up some low end from the chest cavity, but you might also pick up more extraneous noises.

Technique #4: In order to try several mics to see which one will work, set up two microphones at a 45-degree angle to the vocalist so that they make up an equilateral triangle (see Figure 9.49).

You can add a third mic in the center as well, but be sure that the distance is the same as the other two. This also works if you would like to use a combination of microphones.

Technique #5: To get a cool stereo vocal sound, place two condenser mics 1 foot in front of the singer at shoulder height and 2 or 3 feet apart, pointing up toward the mouth. This will yield a kind of wide, thick sound that is very cool if the mix is sparse, but will not do well in a dense mix, as it will tend to sound dark and full. Many singers have trouble with this configuration, so you might have to put up a close dummy mic to sing into.

Technique #6: Using a stereo mic, run one capsule with 10dB more gain on the mic pre than the other. Put a limiter on this one. The one with 10dB more gain should register about 12dB of compression when the singer gets loud. This attenuates the level from the capsule with more gain on it. The net result is, as the singer changes volume, the capsule with the best gain for the application will take precedence.

Technique #7: Many vocalists are just more comfortable with a hand-held mic like they use on stage. Don't be afraid to give them an SM58 if it makes their performance more comfortable and easy. You'd be surprised how good a new 58 can sound with the right preamp.

- **Variation:** If the vocalist wants to sing in the control room, use this same mic but be sure that the vocalist always faces the speakers in order to eliminate the possibility of feedback.

VOICE-OVERS

Voice-overs are recorded far more than most engineers who specialize in music realize, as they're used in radio, television production, filmmaking, theatre, audiobooks, and other industrial presentations. Room requirements for VO work are in some ways less demanding than for a music-recording space. The room should be acoustically dead (not too dead, though) and, most important, really quiet. Nothing will ruin a spot or story quicker than noise leakage on the VO.

Considerations

- The worst thing that can happen on a voice-over is a plosive (breath pop), and therefore it must be avoided at all costs.
- Even among professional voice-over actors, there are some whose voices will sound good on almost any microphone, while others need a specific mic to get the right sound. If you're recording the average person, the variances increase.
- Your mic selection, amount of EQ, and amount of compression used are totally dependent on the voice you're recording.

Placement

Technique #1: The classic way of recording VOs is to place the mic 3 to 4 inches from the talent and off-axis about 45 degrees to prevent popping. Compress about 6 to 9dB at a 4:1 ratio with attack and a fast release (see Figure 9.50).

Figure 9.50: Voice-over using an RE20
© 2023 Bobby Owsinski

Technique #2: Place the mic with the capsule just about even with the talent's nose and point it down at the lips. If popping continues, turn the mic slightly off-axis.

WHISTLING

Whistling is one of the hardest things to record, as there's lots of wind and transients constantly attacking the mic.

Placement

Technique #1: Get the mike off-axis as much as possible, and try a dynamic, such as an SM57. The key is to find the right spot.

Technique #2: Place the microphone by the side of the whistler's head so the mic hears what the whistler hears.

205 | Miking Individual Instruments

10

RECORDING BASIC TRACKS

Basic tracks (sometimes called just basics or tracking) refer to the recording of the rhythm section and are the foundation for any other music overdubs that are to be recorded afterward. That means if there's something faulty in the recording of the basics, it will usually cost time and money to fix it later, which is why it's essential that the basic track recording is as good as it can be, both sound and performance-wise.

Basic tracks can encompass any of the following:

- The entire band, regardless of the number of pieces
- The rhythm section only (drums, bass, guitar, keys)
- Drums, bass, and guitar
- Drums, bass, and keyboards
- Drums and bass only
- Drums and keys only
- Drums and guitar only
- Drums only
- Loops or programmed samples by themselves or with another instrument

Most of the above usually have a guide or scratch vocal recorded at the same time as well to at least provide cues to where the various sections of the song occur. While programming the rhythm section might also qualify as a "basic" session, it doesn't require any microphones or a scratch vocal, so we'll leave it out of the discussion for now.

It's also not uncommon for the drums along with several other instruments to play during the basics, with the idea of only capturing a great drum track and then replacing the other instrument tracks with better-sounding or better-performed overdubs.

PREPARING FOR THE SESSION

One of the keys to a successful basic tracking session is the preparation made beforehand, but before you can prepare for the recording you need some essential information. Here's the minimum that you must determine in advance of the session.

This will usually be provided by the producer, artist, or bandleader, and assumes that you're unfamiliar with the act.

- What type of music will be recorded?
- How many songs do you expect to record?
- Who are the musicians? (If you know some of them it might affect your setup.)
- Who's the producer (if you're not talking to him or her already)?
- What time does the session begin? *Does that mean the downbeat of recording or when the musicians are expected at the studio to load in?*
- How long do you expect the session to go?
- How many musicians will be playing at once?
- What's the instrumentation?
- How large is the drummer's kit? How many toms will the drummer be using?
- Will the guitarist(s) be using an acoustic or electric?
- What kinds of amps will the guitar player(s) and bass player be using?
- Do any of the players expect to use house gear, such as drums, guitar amps, or keyboards?
- How many cue (headphone) mixes will be required?
- Will there be a scratch vocal tracked at the same time?
- Will they bring any special outboard gear or mics that they'd like to use?
- Will they be tracking to loops?
- Will the band be using a click track?
- Do they require any particular instruments, amps, or effects?
- Will this be tracked to the house computer or producer's?
- If recorded to the house computer, will they bring their own drive?
- Will they also bring a backup drive?

Determining the above before the musicians hit the studio can go a long way to a quick and easy setup and an efficient session.

> **TIP: Don't ask for the setup information too far in advance, since much can change by the day of the session. Getting the info the day before the session is usually sufficient.**

Setting Up A Talkback Mic

One thing that engineers, producers, and musicians all hate during a tracking date is when they find it difficult to communicate with one another. Usually it's easy for the control room to speak with the musicians, but it's not easy to hear the musicians speak to the control room through the open mics that are used on the session, since they're adjusted for the louder playing levels instead of talking. That's why it's essential to set up at least one dedicated talkback mic out in the studio with the players, so you can always hear what's happening on that side of the glass.

The type of mic used really doesn't matter, although an omni set in the middle of the studio can work quite well (see Figure 10.1). Sometimes a second talkback mic is also added in a large studio. In fact, some engineers go so far as to set up a dedicated talkback mic for each musician if it's only the four-piece rhythm section recording.

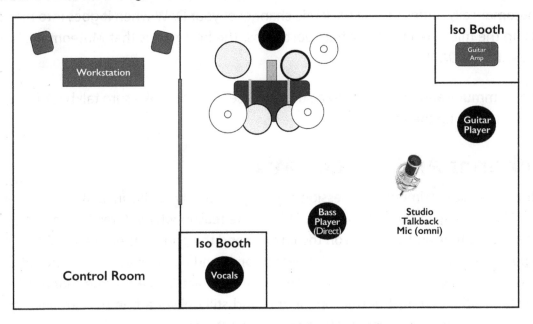

Figure 10.1: A talkback mic set up during a tracking date
© 2023 Bobby Owsinski (Source: iStock Photo)

Regardless of how many mics you use, the talkback mic will make communication between the control room and the studio a lot easier and will keep the musicians a lot happier as a result.

> **TIP: Make sure to mute the talkback mic when the band is playing. It will probably sound tremendously trashy and distorted, since it's set up for people talking and not playing.**

The Importance Of Communication

Not only do players hate it when they're speaking to you from the studio and you can't hear them, but also when they're in the studio and there are long periods of silence between takes as well. They might see a conversation going on, but if they can't hear it, they can get insecure and feel isolated. You may be having a conversation about what kind of take-out food to order, but as far as the player can tell, you're talking about how bad the player's performance was and how you'd like to replace him or her. Get rid of the insecurity by latching the control room talkback so they can hear you all the time between takes.

That said, one of the big problems with most studios built around a DAW is that latching the control room talkback mic and opening up a studio talkback channel isn't as easy as in the days where recording was centered around a console. On many DAWs, this simple action might require multiple selections that have you mousing all over the screen, which sometimes isn't very logical and takes way too much time to execute.

There's a very simple and inexpensive way around this problem though. SoundRadix has created a plugin called Muteomatic that automates the activation of a talkback or listen microphone channel according to your DAW's transport state and then displays the channel's activation status. In other words, it automatically mutes the talkback mic channel of your DAW when it goes into play or record, and opens up the mic when the DAW transport stops. The best part is that Muteomatic is *free* and is compatible with any DAW.

Once again, communication is the key to a successful session, so make sure talkback is high on your agenda when setting up the session.

Headphones And The Cue Mix

Perhaps the greatest detriment to a session running smoothly is the inability for players to hear themselves comfortably in the headphones. This is one reason why veteran engineers spend what seems to be a lot of time and attention on the cue mix and the phones themselves. In fact, a sure sign of an inexperienced engineer is treating the headphones and cue mix as an afterthought instead of spending as much time as required to make them sound great. While it's true that a veteran studio player can shrug off a bad or distorted phone mix and still deliver a fine performance, good "cans" makes a session go faster and easier, and take a variable out of the equation that can sometimes be the biggest detriment to a session.

Whereas at one time each studio had to jerry-rig together their own headphone amp to power their cue mixes, these days it's easy and fairly inexpensive to buy a dedicated headphone system that's easy to set up and sounds great. Companies such as Behringer, Furman, PreSonus, Rolls, and Aphex all make units that will work better and can be a lot cheaper than the traditional method of a large power amp and resistors (see Figure 10.2).

Figure 10.2: PreSonus HP60 headphone amplifier
Courtesy of PreSonus Audio Electronics.

Personal Headphone Mixes

Perhaps the best thing to come along in recent years has been the introduction of the relatively inexpensive "more me" personal headphone systems. These systems allow the musician to build a headphone mix by supplying up to 16 channels of separate instruments and vocals to control. Each headphone mixer/box also contains a headphone amplifier that can (depending upon the product) provide earsplitting level. Manufacturers include Allen & Heath, Roland, Aviom, and Hear Technologies (see Figure 10.3). As above, it's best to provide a stereo monitor mix (what you're listening to) as well as kick, snare, vocal, and whatever other instruments are pertinent.

> **TIPS FOR GREAT CUE MIXES:**
> - Long before the session begins, test every headphone to make sure there's no distortion and they're working correctly (test with actual music).
> - Make sure there's plenty of headphone cable available so that the musicians can move around as needed. Use cable extenders as necessary.
> - Check to make sure that the cables are not intermittent (nothing stops a session as fast as a crackling headphone).
> - Some engineers send the stereo monitor mix (the mix that you're listening to in the control room) to the phones first and then add a little of the individual instruments as needed (called "more me"). This is a lot easier than building up individual mixes, unless that's what the musicians request.
> - Many studio cue systems allow the musicians to build their.

Figure 10.3: Hear Back Pro 4 Pack headphone system
Courtesy of Hear Technologies

The Click Track

The click track, or recording to a metronome, has become a fact of life in most recording because not only does playing at an even tempo sound better, but it also makes cut-and-paste editing between performances in a DAW possible. Playing to a click can present a number of problems, however, such as leakage of the click into the mics, and the fact that some people just can't play in time to save their lives. We'll cover these shortly.

Making The Click Cut Through The Mix

Many times just providing a metronome in the phones isn't enough. What good is a click if you can't hear it or, worse yet, can't groove to it? Here are some tricks not only to make the click listenable, but also to make it cut through the densest mixes and seem like another instrument in the track.

- **Pick the right sound.** Something that's more musical than an electronic click is better to groove to. Try either a cowbell, a sidestick, or even a conga slap. Needless to say, when you pick a sound to replace the click, it should fit with the context of the song. Many drummers like two sounds for the click-something like a high go-go bell for the downbeat and a low go-go bell for the other beats, or vice versa.

- **Pick the right number of clicks per bar.** Some players like quarter notes while others play a lot better with eighths. Whichever it is, it will work better if there's more emphasis on the downbeat (beat 1) than on the others.

- **Make it groove.** Adding a little delay to the sound can make it swing a bit and not sound as stiff. This makes it easier for players who normally have trouble playing to a click. As a side benefit, this can help make any bleed that does occur less offensive, as it will seem like part of the song.

Preventing Click Bleed

Now that the click cuts through the mix, it might do it so well that it starts bleeding into the mics (especially drum mics). If that's the case, try the following:

- **Change to a different headphone.** Try a pair that has a better seal. The Sony 7506 phones provide a fairly good seal, but a good pair of isolation headphones like the Metrophones Studio Kans or the Vic Firth SIH2s (see Figure 10.4) will isolate a click from bleeding into nearby mics.

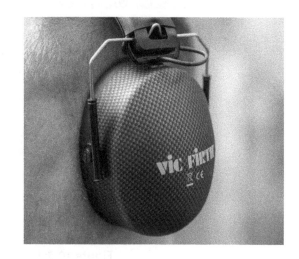

Figure 10.4: Vic Firth SIH2 isolation headphones
Courtesy of Vic Firth Company

- **Run the click through an equalizer** and roll off the high end just enough to cut down on the bleed.

- **Have the players use one-eared headphones,** such as the Stanton DJ Pro 300 or the Beyerdynamic DT 1S. Many times players will leave the phones loose so they can hear what's going on in the room. If they can have a click in one ear in the headphone that's sealed closely to the head, they can get the live room sound in their free ear.
- **Run the click to just one person** (such as the drummer or the conductor) and let that player communicate the click to the band through a solid performance.

If all else fails, try this method. It might even provide the loosest feel and best groove, too.

1. Record the song three times with the click.
2. Choose the best version.
3. Instead of a click, use the track for the drummer to play against by muting the drum part and just playing back the other instruments.
4. Proceed with recording overdubs.

Using the above method, the drummer can hear the rest of the band and play along through headphones so that there should be very little bleed. Once the drums are printed, the session can progress as normal.

When A Click Won't Work

Let's face it, there are some musicians who don't like to play to a click. It's unnatural and doesn't breathe like real players do. That said, in this world of drum machines, sequencers, loops, and DAWs, most musicians have grown used to playing against a metronome.

There are still those times and those players (and it's usually the drummer) where a click just won't work. The performance suffers so much that you get something that's not worth recording. No problem. Don't get obsessed with the click or the fact that the tempo fluctuates without it. You'd be surprised by the number of great hits that have been recorded without a click and have wavering tempos, especially back before the '80s. Remember, feel and vibe are what make the track, not perfection.

THE TRACKING SESSION

While many modern recordings are made with as few musicians playing at once as possible, most recording veterans prefer to have as many players as possible during the basic tracking date. The reasons? The vibe and the sound. While such a session can be rather nerve-wracking in complexity, it can be a lot of fun as well.

Where To Place The Players In The Room

Regardless of how good the headphone system is, the players won't play their best unless they can see each other. Players in the studio react to the nuances in a song by looking at other players, so a clean sight-line to each player is a must. You'll also find that many players (especially studio veterans) rely on looking at the drummer playing the snare to stay locked in time. In fact, a trick that many bass players use is to place their foot up against the bass drum peg so they can feel the pulse of the drum.

Since communication is essential during a session, it's also important to set up the players close enough to one another that they can talk without the benefit of their headphones (see Figure 10.5). There's so much that players need to discuss, especially when running down a song for the first time, that close proximity to one another can go a long way toward an efficient session with great performances.

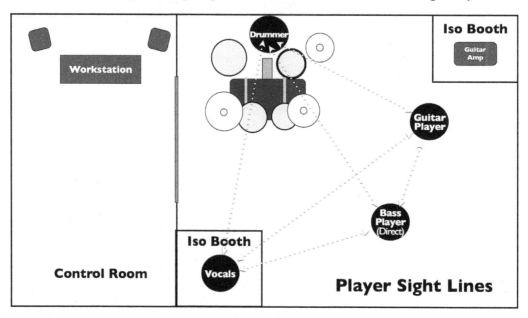

Figure 10.5: Player sight-lines
© 2023 Bobby Owsinski (Source: iStock Photo)

How Long Should It Take?

Generally speaking, if the session is on a budget and needs to be finished in a day, you should aim to be recording within the first hour after the musicians arrive and are set up. Provided that you're already set up by the time the band arrives, the time it takes to get recording will be cut to a minimum.

It all depends on the budget allocated to the project though. On a typical album-tracking session that will span a number of days or even weeks, the first day of tracking may be used as just a setup day. In that case it may take half the day for everyone to feel comfortable, for the engineer to get sounds, and for the musicians to get their headphone mixes together. Somewhere during the second half of the day is when the band actually begins recording.

There was a time during the '70s when a few high-budget projects would take an entire week just to get the snare drum sound. While they might've attained temporary snare drum nirvana, 99.99999 percent of sessions have to move faster than that, and they should. The more time you take before recording, the less time you'll actually spend recording, since the players' attention spans decrease proportionately. Although you want things to sound as good as possible, a poor-sounding track with a great vibe is a lot more usable than a well-recorded but musically stale track.

> TIP: The idea is to be making music quickly with everybody hearing themselves. Get something recorded as soon as possible and have the musicians hear a playback. If any sonic adjustments are required, good musicians will hear what needs to be done and usually do it themselves.

Recording Without Headphones

Many bands not used to recording in a studio may not adjust well to using headphones. Here's a way that allows them to feel comfortable while recording without having to put tiny speakers on their heads.

- **Get the players set up close to the drums.** This helps to minimize the acoustic-delay times that can occur when you spread the players out too far, and is most likely what they're used to when playing live or rehearsing.

- **If there are two guitar players, set them up on opposite sides of the kit.** This will provide a better stereo picture for the inevitable leakage that will occur. (See the next section.)

- **A floor monitor can help the vocalist sing a scratch vocal.** Placing the singer in the same room with the band will get a better performance because that's what they're used to.

- **Most of the time the singer will actually gravitate to the spot in the room where the band's balance is best.** That's where you have to move the floor monitor to.

Leakage

Many inexperienced engineers sometimes go to great lengths to avoid the acoustic spill known as leakage, where the sound from one instrument leaks into another's mic. True isolation is actually pretty difficult to achieve when tracking multiple instruments, contrary to what many believe. You may be surprised to learn that many A-list engineers don't view leakage as undesirable at all, preferring instead to embrace it and use it to embellish the track (see the interviews in Part II).

The fact is that leakage can sometimes make the overall sound much bigger, especially if you're recording in a good-sounding room. Leakage can be used as a sort of glue between instruments in much the same way that instruments magnify one another in a live situation. Remember that all of those great hits from the '50s, '60s, and early '70s had a ton of leakage on them, and some of those records are still considered some of the best-sounding ones ever.

So, when tracking with multiple instruments, keep the following in mind:

- Keep the players as close together as possible. Not only will it help the players communicate, but the leakage will contain more direct sound than room reflections, which will sound better.

- Whenever possible, use omnidirectional pattern microphones. The leakage picked up by omnis tends to be a lot less colored than directional microphones.

> **TIP: There's a difference between good- and bad-sounding leakage. Good mics usually provide good-sounding leakage.**

A Couple Of Non-Essentials

Here are a couple of tips for tracking that, while not exactly essential to the tracking process, can sure make your life easier when it comes time for overdubs.

Recording A Tuning Reference

Record a 10-second tuning note from the instrument in the session with the most stable tuning. This would be an instrument whose tuning doesn't move, such as a synth or an organ, or in the case of a guitar-only band, the guitar that's regarded as most in-tune. If your tuning reference is guitar only, play a single A note as well as A and E chords.

The reason why you might want to do this is that if for some reason the band happened to use a tuning that was a couple of cents flat (like when using an acoustic piano that wasn't tuned for the recording), they could easily use the tuning note as their reference later during overdubs to get in tune. This seems like such a small thing, but you wouldn't believe how much time it can save you down the road if a situation arises where you just can't figure out why everything sounds out of tune.

> **TIP: Have all instruments use the same tuner if there seems to be a tuning problem.**

Recording A Count-Off

A recorded count-off is important for those times when an overdub is required before the song starts. Even if you're playing to a click that's being generated by the DAW itself, recording the click at least four bars ahead of the downbeat is a foolproof way to make sure that any pickup or opening part is easily executed.

If a click isn't being used, it's even more important to record the count-off. Have the drummer click two bars before the count with his or her drumsticks and then count, "One, (click), Two, (click), One, Two, Three, (silent click)." Sometimes a count with the last two beats silent is used instead, like this; "One, (click), Two, (click), One, Two, (silent click), (silent click)." This is plenty of count for the band to get the feel, and the silent clicks on the end makes it easy to edit out later without worrying about clipping the downbeat.

Recording An Electric Guitar Direct Signal

While this isn't something that was done much in the past, today it's always a good idea to take a direct signal from an electric guitar as well as miking the amplifier. There are two reasons for this:

- It's a lot easier to edit a clean guitar waveform than a distorted one.
- You might want to change the sound later and a clean version of the performance provides many more possibilities thanks to amplifier emulation plugins or reamping.

Remember that just because you've recorded that clean direct version doesn't mean you have to use it. It's just there to make your life easier as the project progresses.

HOME STUDIO RECORDING TECHNIQUES

As you've read, sessions involving bands or session players revolve around the basic tracking session, which is critical to the feel of the song and any parts that are added to the project down the line. Many producers and songwriters don't work like that though, especially if they're working alone in their home studio.

A modern pop song might start out based on an interesting loop or sample, with the rest of the song built for there. That might mean adding additional loops or real player overdubs, but it's not unheard of for a songwriting session to turn into the final track with no demo stage in-between.

Also, because fewer musicians or vocalists are used, it's not uncommon for a song to go from conception to finished product in less than a day. Needless to say, this is mind-blowing to old-school musicians, producers and engineers.

As a result, home studio recording can be a hybrid of songwriting, tracking, overdubs and mixing happening simultaneously or in quick succession. While that's the best case scenario, things normally go slower than that, especially if you're new to recording and still getting used to how everything works and sounds.

The point is, many of the time-frames used in this chapter pertain to seasoned pros who know what they're looking for and how to get it. There are just as many who will be meticulous in finding sounds, parts and performances that will take a lot longer, and some that will never be satisfied and therefore never finish.

A good rule of thumb is, take as much time as you need to get the sounds that you want and learn what you need to, but it's better to finish something and have it released than to try for perfection and never release anything.

PROCESSING DURING RECORDING

The choice of whether to add any processing, such as EQ or compression, during recording is made on an individual track basis. The old adage applies: "If it sounds right, it is right." However, many engineers would rather take the time to position a mic correctly before reaching for an EQ or compressor, keeping any processing during recording to a bare minimum. Also, remember that if you record with compression or EQ, that processing can't be removed after the fact.

If you choose to record with compression, be careful not to set it so that it squashes the important transient attacks of the the instrument you're trying to capture. Depending on how an instrument is played, it can have quite a wide range of dynamics, from extremely quiet to harshly aggressive. Compression needs to be set based upon the intensity of the sound at the mic during recording. Sometimes just a touch of compression can help bring even out the differences in note volumes. Many times an engineer will add a limiter to make sure than any wild transients are caught before they overload the recording.

If you are going to record with compression or limiting, try using a ratio of around 3:1 or 4:1 with a medium/fast attack and a relatively fast release. Also, set the threshold so the loudest peaks have no more than a dB or two of gain reduction. If you feel that you don't need any compression, don't use it!

With EQ, the same rule applies. If you don't need it, don't use it. That being said, engineers will often apply a high-pass filter on the preamp or the mic itself to help reduce any low-end rumble. If you feel the need to add EQ during tracking, try moving the mic first, or try another mic. Remember that EQ can always be added later during the mix.

On the other hand, many experienced engineers have no fear of recording with EQ and compression as they want the tracks to sound as close to a finished record as soon as they can (see the interviews in Part 2 for some examples). This comes after years of experience and experimentation, so they feel comfortable that any processing choices they make are correct and they won't rethink them later in the project.

EQing The Microphones

I once did a tracking session with the legendary engineer/producer Ken Scott (The Beatles, David Bowie, Supertramp, Duran Duran) where we needed to change out the the drum kit for another one. After the new kit was in place, Ken positioned the microphones and jumped right into recording without touching the EQ on any of them.

It was at that point that I realized that he was EQing the microphones and not the instruments. He was so familiar with the microphones (he used the same ones for decades) that he knew exactly what their deficiencies were, and so he EQed them accordingly. This technique is not practiced very often, but he can be a very efficient way of recording, provided that you use the same gear every time and have plenty of experience with it.

THE ASSISTANT ENGINEER

Many musicians that make the jump to engineer start out as assistants. Once that meant that you started as a runner (a person that makes coffee, takes food orders, and cleans up) at a commercial studio facility and eventually moved up to assistant engineer after you proved your reliability and learned the ropes. The advantage of working in that type of studio was that you got an opportunity to watch and learn from a variety of producers and engineers, and discover your personal reference point on what sounds good and what doesn't.

Today there are actually two types of assistant engineers - the traditional commercial studio job, and working for a producer in his or her personal studio. There are fewer commercial studios and more producers working at home, but either is a pathway to furthering your career as an engineer or producer.

Expectations

Regardless of who you work for, the expectations of an assistant are pretty much the same. The job today requires a large amount of computer literacy, and many of the every day tasks include:

- Creating a session
- Creating session backups
- Setting permissions
- Clocking external machines and gear
- IP addressing
- Fixing crossfades
- Editing out pops and clicks
- Importing tracks and stems
- Clip-gaining tracks and stems
- Finding missing parts and plugins
- Tuning vocals or instruments
- Updating software and firmware
- Troubleshooting computer issues
- and dozens of other computer-related concerns that pop up

Among the traditional tasks still expected of assistants are:

- Prepping the room for a session
- Setting up microphones
- Setting up cue mixes
- Patching in outboard gear
- Taking food orders
- Making coffee

While you may never work in a studio that has a second engineer, and if you own your own gear, you may never be one yourself, it's good to know what an assistant in a major facility such as the Record Plant, Capitol, Ocean Way, or Power Station really needs to know. Some of these tips come from the legendary Al Schmitt (who's won more Grammys than any other engineer), and they will help you understand what's expected of an assistant and how to run a professional session, regardless of the level that you're on.

1. **Good assistants know every piece of gear in the studio inside and out.** This is especially helpful to incoming independent engineers who've never worked in the room before.

2. **Good assistants know the acoustics of all the live spaces.** They're able to suggest to the visiting engineer which instruments record best in a particular place in the room.

3. **Good assistants know every mic in the studio very well.** They know which pair of mics match, which mics work best on certain instruments, and which mics have a problem.

4. **Good assistants are well-versed in various DAWs.** While Pro Tools may be the standard DAW used in every major recording and post studio, today an assistant has to be versed in other DAWs like Logic, Cubase and Ableton as well. Most assistants will be in charge of running the DAW, and they are better at it than everyone else in the session.

5. **Good personal hygiene is a must.** No one likes to be in a room with someone who has body odor or bad breath, and artists and producers won't put up with it. Take a bath, put on clean clothes every day, and keep the breath mints handy if you want to keep your job.

6. **Good assistants are transparent.** When you need them, they're there; when they're not needed, they're in the background. A good assistant is always seen but not heard, and never offers an opinion even when asked. A great attitude is a prerequisite for the job, as well as leaving your ego at the door.

7. **Good assistants admit mistakes.** If you make a mistake, admit it as soon as possible. You may have to take your lumps, but we'll fix it and move on.

8. **Good assistants don't guess.** If someone asks you something that you don't know, be honest and don't guess. There are plenty of ways to find out something in a hurry if you don't know right now.

9. **Good assistants keep a notebook.** They keep track of all the details of the session, from the setup, to the players, to the mics used, to which songs were recorded in what order, to everything else. It's a great learning tool, but it may also come in handy later in the project or on the next one.

10. **Good assistants know how to make coffee.** Coffee is still the fuel that powers a recording session. The better the coffee, the happier everyone will be.

11

RECORDING OVERDUBS

The overdub session has truly evolved over the years as track counts have grown. Whereas once a normal recording required either zero or just a few overdubs, The Beatles took the overdub session to a whole new level even during the 4-track days, using it as a creative tool to enhance the tracks they started with. As the track count grew, so did the purpose of the overdub, with many acts using it more as a way to perfect each part and maintain isolation between parts during mixing. Today, the enormous number of tracks available to us has changed the role of the overdub into one of alternatives, where every sound and idea can be recorded without having to erase anything first.

That ability to record everything can actually be somewhat of a curse, because capturing every idea sometimes gets in the way of making a decision and sticking with it, as was required in the old days limited by the few tracks that were available. If there's one thing that old-school engineers universally rail against, it's that inability to make a decision rather than piling everything but the kitchen sink into the session with the hope of sorting it out later during the mix.

OVERDUB SETUP

Having almost unlimited tracks also lends itself to a whole host of production techniques, such as doubling and layering. While these specific techniques go beyond the scope of this book (see The Mixing Engineer's Handbook 5th edition for more about production and mixing techniques), there are a number of basic points to consider before beginning overdubs.

Recording In The Control Room

Regardless of the instrument, it's always best if you can get the player to record in the control room with you (if you have one). This is easy with guitar, bass, electronic keys, and even vocals, and it's tougher for everything else, but having the ability to instantly communicate with the player and have him hear what you're hearing often raises the performance level.

For players with electric instruments that require an amp, cables and hardware are now widely available to keep the amp or speaker cabinet in another room while the player plays in the room with you, although some prefer to be near their speaker cabinet for feedback effects. That said, playing in the control room is sometimes not an option if more than one player at a time is required, or with instruments that are quiet such as some percussion, acoustic guitars, or strings. Recording vocals in the control room was covered in the vocal section of Chapter 9.

Setting Up The Overdub In The Studio

It's not always possible or desirable to have the player in the control room area with you, but there are a few things to consider when doing overdubs in the studio.

If you're doing fixes to basic tracks, keep the instrument or amp in the exact same spot as it was during the tracking session. In fact, try not to move any of the other instruments as well. You'd like to keep the room reflections exactly the same so the sound will match.

If you're doing an overdub that has no relation to the basic track, move the vocal or instrument into the big part or best sounding part of the studio. All instruments sound best when there's some space around them for the sound to develop.

TIP: You can cut down on any unwanted reflections from the room by placing baffles (sometimes called gobos or screens) around the mic, the player, or the singer as illustrated in Figure 11.1.

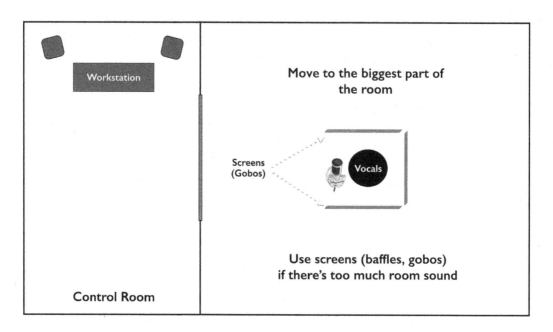

Figure 11.1: Placing screens around the vocalist
© 2023 Bobby Owsinski (Source: iStock Photo)

OVERDUBBING INDIVIDUAL INSTRUMENTS

While we covered the majority of recording considerations in Chapter 9, there are a few areas to look at when it comes to overdubbing the four most frequently overdubbed instruments: vocals, keyboards, acoustic guitar, and electric guitar.

Getting the Most From A Vocalist

One of the hardest things to record can be a vocalist who is uncomfortable. Even a seasoned pro sometimes can't do her best unless the conditions are right. Consider some of these suggestions before and during a vocal session.

- Make sure the lighting is correct. Most vocalists prefer the lights lower when singing.
- A touch of reverb or delay for some ambiance in the headphones can be helpful.
- If you need to have the singer sing harder, louder, or more aggressively, turn down the vocal track in the phones or turn up the backing tracks.
- If you need to have the singer sing softer or more intimately, turn up the singer's track in the phones or turn down the backing tracks.
- Keep talking with the artist between takes. Leave the talkback engaged if possible. Long periods of silence from the control room are a mood killer.
- Try turning off the lights in the control room so the singer can't see you. Some people think that you're in there judging them when you might be talking about something completely different.
- If the take wasn't good for whatever reason, start by telling the vocalist what was good about their performance, then explain what was wrong in a kind and gentle way - something like, "That was really good, but I think you can do it even better. The pitch was off a little during the third phrase." This goes for just about any overdub, since players generally like to know what was wrong with the take rather than be given a "Do it again" blanket statement.

TIP: With any overdub, it's important to always keep smiling. Virtually all players become insecure at some point when under the microscope of the studio, and the slightest negative, even though subliminal, can multiply in the player's head in no time. A smile goes a long way to adding to the player's comfort level.

The 3 P's

My good buddy and fantastic engineer/producer Ed Seay has a saying that I've used in a number of books about the 3 P's or "Pitch, Passion and Pocket." That refers to the 3 things that every great vocal must have, although you can apply it to other musical performances as well. Ed is not only one of the greatest mixers anywhere, but he's a great teacher and mentor too, having brought along production luminaries such as Dave Pensado and Brendon O'Brian, not to mentioned having worked on tons of hits by Alabama, Martina McBride, Dolly Parton, and many more, so he knows what he's talking about.

Here's an excerpt that explains Ed's "Pitch, Passion and Pocket" concept from The Mixing Engineer's Handbook (Ed's interviewed in it).

"In the studio, the 3 P's are what a producer lives by. You've got to have all three to have a dynamite vocal. And while Pitch and Pocket problems can be fixed by studio trickery, if you don't have Passion, you don't have a vocal. On stage, the three P's apply maybe even more so, since you don't have any of the advantages of the studio. Let's take a look inside the three P's.

Pitch

Staying in pitch means singing in tune. And not just some of the notes – every single note! They're all equally important!! Pitch also means following the melody reliably. There's a trend these days to skat sing around a melody, and while that might be desirable in some genres, it doesn't work in any genre if you do it all the time. Skating might show off your technique and ability but a song has a melody for a reason. That's what people know, that's what they can sing to themselves, and usually that's what they want to hear.

Pocket

The Pocket means singing in time and in the "groove" (the rhythm) of the song. You can be in pitch, but if you're wavering ahead or behind the beat it won't feel right. The vocalist must concentrate on the downbeat (on beat 1) to get the entrances and on the snare drum (on 2 and 4) to stay in the pocket.

Passion

Passion is not necessarily something that can be taught. To some degree, you either have it or you don't. What is Passion? It's the ability to sell the lyrical content of the song through performance. It's the ability to make me believe in what you're singing, that you're talking directly to me and not anyone else. And passion can sometimes trump pitch and pocket. A not-all-that-great singer who can convey the emotion in his or her voice is way more interesting to listen to than a polished singer who hits every note perfectly but with little emotion. In fact, just about any vocalist you'd consider a "star" has passion, and that's why he or she is a star.

On-stage, Passion can sometimes take a back seat to stamina, since you have to save yourself for a whole show and you can't blow it all out in one song. That's why many singers have only one or two big "production numbers" where they totally whip it out. This means that you have to learn the limits of your voice, learn how much of you goes into just cruising and when you can do it, and how much you need left in the tank to do your biggest, most effective show stoppers.

In the studio, there's never any cruising – you've got to give all the passion you can give for every song. A few paragraphs ago I said that you either have passion or you don't, but sometimes you really have it and you don't know it, and it's the job of the producer to pull it out of you. That could mean getting the singer angry to stir some emotion, building him or her up by telling him how good he is, or making her laugh to loosen her up. Anything to sell the song! But once you know how to summon it up from inside you, you can do it again and again.

In the end it's simple. Even though the singer has to create the performance, it's up to producer to extract the 3 P's."

Electric And Virtual Keyboard Overdubs

Virtually all electric and virtual keyboards have a stereo output, but in the majority of cases the stereo is derived from taking a mono signal and blending it with an effect such as a chorus. While this can sound great on its own, adding a few layers of keyboards to a track can soon make this effect turn into something called big mono (see Figure 11.2).

The more keyboard layers that are added and the wider they're panned, the more the individual parts blend together-more because of the effect used to stereoize them than anything else.

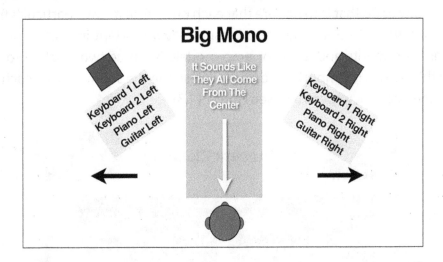

Figure 11.2: Big mono
© 2023 Bobby Owsinski

While this blending might work in some cases, in others you probably want to hear each of these parts distinctly from one another. There are three ways to overcome big mono:

1. Record the instrument in mono by using the output labeled "Mono," which usually is a dry signal. You can stereoize it in the mix with effects that are more appropriate for the song.
2. Some keyboards have mono samples just for recording, and these are the ones to use.
3. A few keyboards that use samples use samples that were recorded in true stereo. Even if you use the stereo sample, try to pan it to a section of the stereo soundfield other than hard right and hard left.

> **TIP: If you must record in stereo with the effect, you can minimize the big-mono effect by making sure that no two instruments are panned on top of one another (such as hard left and hard right) and instead each stereo instrument is panned to its own space in the stereo soundfield.**

Electric Guitar Overdubs

The electric guitar has a huge impact on most modern music today, but just capturing its sound sometimes isn't enough. Since many songs feature either multiple guitar players or overdubs, each part has to be able to fit seamlessly with the other, as well as with the rest of the track. Here are a number of ideas for how that's frequently accomplished.

Tone Controls

Many players and engineers alike overlook the effect that the tone controls on the amp, guitar, or plugin can have on how the guitar part fits in a song. You want to be sure that every instrument is distinctly heard, and the only way to do that is to be sure that each one sits in its own particular frequency range, and the tone controls will help shape this. This is especially important in a situation with two guitar parts using similar instruments and amps or amp models (such as two Strats through two Fender Super Reverbs). If this occurs, it's important to be able to shape the sound so that each guitar occupies a different part of the frequency spectrum (see Figure 11.3).

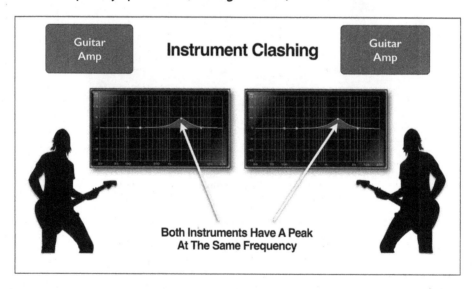

Figure 11.3: Instrument frequency clashing
© 2023 Bobby Owsinski (Source: iStock Photo)

To make our example work in the mix, one guitar would occupy more of a higher-frequency register, while the other would be in a lower register, which would mean that one guitar has more high end while the second guitar is fatter-sounding, or both guitars might have different midrange peaks. Another way to achieve the same thing is for one of the guitars to roll back the guitar tone control (although most players hate to do that).

> **TIP: Remember that it's not how an instrument sounds by itself but how it sounds in the mix with other mix elements. A instrument that sounds good on its own won't necessarily work when mixed in the track.**

Instrument Variety

An easier way to get a variety of sounds is by using different guitars and amplifiers during the recording, which is why studio musicians usually bring such a huge assortment of gear to a session. Although you can change the sound of your main axe by selecting a different pickup, changing to a different amp channel, or using a different pedal, sometimes that's really not enough of a meaningful difference in the context of the song.

The ideal situation is for a player to have at least a guitar with humbucking pickups and another with single-coil pickups, and a variety of amp sounds from Marshall-like to Fender-like to even Vox-like, which are all distinctly different (easy to do with today's amp simulators). This variety develops a varied sonic landscape that is more interesting to the ear. If each guitar track can sound somewhat different, that means the mixer won't need to spend as much time trying to do the same thing with processing later (even if you're the mixer).

Amp Variety

In an odd paradox, smaller amps and speakers tend to sometimes sound bigger than large amps and speakers when recording. One of the reasons is that a 12-inch cone at low volume has a relatively shallow cone excursion (its travel back and forth), while the same voltage from an amp to an 8-inch speaker will produce a larger cone excursion, causing it to pump more air.

That's why it's not uncommon for players in the studio to power small speakers (such as 8 and 10 inches) with a larger amp (such as 50 to 100 watts). For an overdriven sound that requires you to crank it to 10, small speakers are going to pop in no time, and you're better off with your normal amp and cabinet combination, but if you're just looking for some edge to a clean sound, a large amp with a small speaker is the way to go in the studio.

Another combination that's frequently used is a small-wattage amp, such as a 5 watt Fender Custom Champ or 20 watt Orange Micro Terror and a large speaker cabinet like a Marshall 4x12. As long as the impedance of the cabinet matches the impedance that the amp is looking for, it's hard to believe how big the sound can be as a result.

> **TIP: A small-wattage amp with a larger speaker or speakers allows you to turn the amp up to the point where it's overdriven, but without blowing down the studio walls. It's an effective technique for getting some sounds that you might not get any other way.**

Going Direct

While many of the newer amplifiers have outputs intended for direct recording, you might not get the same sound as when miking the speaker cabinet. That's because in most cases you're hearing the sound of only the preamp section of the amplifier, which sounds nothing like an amp cranked through the speakers. If you treat this sound as just a slightly more effected clean direct signal, however, you might find it very pleasing to work with.

Another way to record direct is to feed a signal from the extension speaker jack of the amp into a direct box that has the ability to accept this type of input (see Figure 11.4). Usually the DI will have two inputs: one labeled Guitar and the other labeled Amp or Speaker. Make sure that you only connect to the Amp or Speaker input, as the voltage coming from the extension speaker output is high enough to destroy the direct box if plugged into the Guitar input and may even damage the amp as well in rare cases. As with the direct output from the amp, the sound will not be what you experience out of the speakers, so you may have to adjust the amp's controls in order to get a sound that you find useful.

TIP: Keep in mind that the settings you normally use on the amp might have to be changed in order to get a usable direct sound.

Figure 11.4: A direct box Speaker switch
© 2023 Bobby Owsinski

Using The Direct Signal For Editing

Distorted electric guitar is notoriously difficult to edit, since it's difficult to see the attacks and releases of notes and phrases. A great way to make it easier to see the potential edit points is to always record a direct track along with the amplifier mic or amp emulator track. This track may never be used in the final mix, but will more easily show the natural edit points of the track (see Figure 11.5). This also makes it easy to change the sound with a plugin later to better fit the track.

TIP: After editing, be sure to hide and disable the direct edit track to unclutter the mix window and free up system resources.

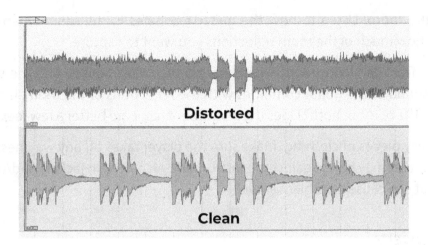

Figure 11.5: A distorted and clean edit track
© 2023 Bobby Owsinski

Using An Amplifier Emulator

Direct guitar recording is no longer a big deal because there are so many effects boxes and amplifier emulators on the market that are capable of acting as a sort of "super direct box" for recording. Starting with Line 6's Pod in 1998, just about every manufacturer now offers an inexpensive amplifier emulator capable of direct recording.

Regardless of which unit you use, keep the following in mind:

- **Be judicious with the distortion and sustain.** Lots of distortion and sustain is fun to play with, but it isn't always appropriate for the song. Be prepared to dial it back to make the part fit better in the mix, especially if there will be other guitar parts added later.

- **Be judicious with the effects.** One of the cool things about modeling and multi-effects boxes is that you can get such a wide variety of sounds, many with over-the-top effects. Just like with distortion, think of what's appropriate for the song, not what feels fun to play with. Once again, take into account how everything will fit together in the mix, especially if you add additional parts.

Acoustic Guitar Overdubs

When it comes to overdubbing acoustic guitar, there's far more to it than meets the eye if you want to get a great recording. Here are a number of considerations for that overdub.

Acoustic Guitar Recording Preparation

Acoustic guitar recording requires a preparation that's different from just about any other instrument. Before you begin recording even the first note, there are a few steps you should ask the player to take.

- **Change the strings.** Putting on a fresh set of strings will not only help with the tuning, but make the instrument resonate better as well.

- **Listen to the room.** Listen to how the guitar resonates in the room when played. That will determine how much of the room reflections you want to capture.
- **Stand back from the instrument.** Move around the instrument to find the sweet spot where the direct sound of the guitar combines with the reflections of the room. Is the best sound at the neck, at the body, or both? Does the instrument resonate better a few feet back or up close?
- **Take off noisy pieces of clothing.** Make sure the player takes off any watches, rings, jewelry, or belt buckles that may bang against the instrument. Also, certain jackets and/or shirts may have buttons that can cause a problem as well.

Instrument Variety

You can have the world's best-sounding acoustic guitar, but if it doesn't fit in the track, it's nice to have an alternative available instead of reaching for the EQ. Even more so than electric guitars, a variety of acoustics will make it a lot easier to dial in the right sound for the track. After all, changing the sound of an electric is pretty easy by selecting a different pickup, amp channel, or amp model, but what you see is what you get with an acoustic.

Ideally, you'd want an acoustic that's bright-sounding and another that's darker. You can achieve this by having either guitars with different body styles or guitars made from different woods. Sometimes even having the exact body types with the same wood can still yield a substantially different sound, depending upon the age and make of the guitar. So before you begin recording, cover your bases and ask the player to borrow or rent several different acoustics for the session.

TIP: Don't discount inexpensive acoustic guitars when it comes to recording. Sometimes a cheap acoustic that sounds horrible on its own can be magic in the track.

Going Direct

While just about any good preamp will get the job done, many engineers tend to have a favorite model they turn to when recording acoustic guitars that have a pickup. Just to be clear, we're talking about a microphone preamp that connects directly to your recorder and not the kind of DI or preamp that's expressly designed for an acoustic, like an LR Baggs Para Acoustic DI. To properly capture the lows, mids, and highs of an acoustic guitar, many choose to combine a good mic with as neutral a preamp as possible, meaning one that does not impart any sonic color of its own on the guitar.

Since everyone has different ears and opinions, the type of preamp to choose is an open-ended topic, but if you have one that has a "transparent" sound, try that first. If not, try to at least use the highest quality preamp you can.

12

IMMERSIVE RECORDING TECHNIQUES

Although immersive mixing receives all the attention these days, immersive recording isn't often considered. Even though orchestral and field recordists have been recording specifically for 5.1 surround (left front, center front, right front, left and right rear, and subwoofer) for decades, engineers in rock, pop, R&B, and jazz tend to ignore the possibilities. Since the aim of any recording is to capture the sound of the environment as well as the source, immersive miking accomplishes this goal to an extent that we've never heard before. Any of the methods below add a spaciousness that you simply can't even approximate with outboard processors or any other previously mentioned miking techniques.

Today immersive recording can be broken down into three categories: channel-based surround miking, binaural and ambisonics. Let's take a look at the configurations, tools and differences of each category.

CHANNEL-BASED SURROUND RECORDING

Channel-based surround means that the recording is intended for playback for ear-height speakers, with each mic channel intended for a dedicated speaker (i.e. 5.1 miking for 5.1 speaker playback). While there have been some miking setups that contain overhead height channels like in Atmos, this isn't the norm, especially when it comes to normal surround sound playback.

At it's most basic, channel-based surround miking is just an extension of normal stereo miking techniques. Here are a few somewhat standard approaches to consider.

OCT Surround

Optimized Cardioid Triangle (OCT) is a modified Decca Tree (check back to Chapter 6) that uses three cardioid microphones in a triangle with the center mic (C) about 3 inches or so from the center, and the side mics (L and R, which face out towards the sides) 15 to 36 inches away from each other. By adding two additional rear cardioids 15 inches back from the L and R and 8 inches farther outside the L and R and pointing to the rear, you can derive a 5.1 version of OCT (see Figure 12.1). When the L and R cardioid microphones are positioned closer together, the image will sound narrower. Conversely, if the L and R mics are further apart, the image will be wider. For better low-end response, you can substitute mics with omni pickup patterns instead of cardioid.

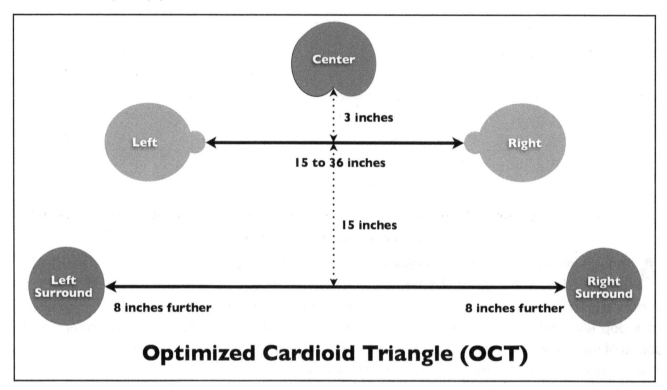

Figure 12.1: OCT surround
© 2023 Bobby Owsinski

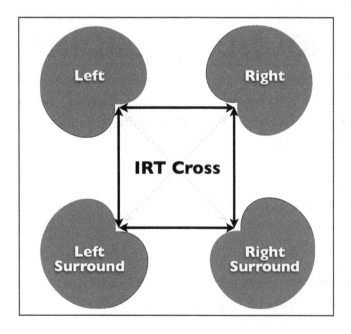

Figure 12.2: The IRT Cross
© 2023 Bobby Owsinski

IRT Cross

This is a configuration that can be used along with one of the stereo miking techniques as a means to capture more ambience, or used by itself. IRT stands for the German-based Institute of Radio Technology, where this technique was created. In essence this is a double-ORTF setup (see ORTF in Chapter 6) with four cardioids arranged in a perfect-square shape with an angle of 90 degrees to each other, respectively. To compensate for the narrower angle compared to ORTF (which is 110 degrees), the distance between the mics is greater (8 inches compared to 6 inches with ORTF).

Strictly speaking, the IRT microphone cross is an array intended for ambiance recording. Its prime characteristic is a transparent and spatial reproduction of the acoustic environment and was used for many years on NPR's "Radio Expeditions" spectacular recordings (see Figure 12.2).

Hamasaki Square

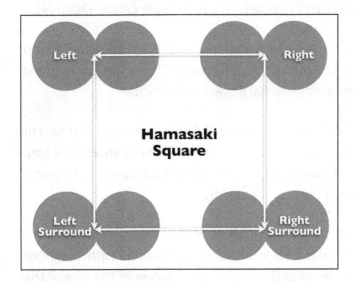

Another configuration intended as an extension of stereo miking, the Hamasaki Square configuration is similar to the IRT cross except that figure-8's are substituted for cardioids. The length of each side is much wider, at about 6 feet, and the figure-8's have their nulls turned to the front so that this array is relatively insensitive to direct sound (see Figure 12.3).

Figure 12.3: The Hamasaki Square
© 2023 Bobby Owsinski

Ideally this array is placed toward the back of the studio or venue and high in the space. The back two microphones are mixed to the surround channels, while the front two channels are combined with the front left and front right signals. This technique is less sensitive to the distance between the main array and the ambience array than other ambience techniques.

Double M-S

This method uses a standard M-S miking configuration with the addition of a rear-facing cardioid mic (see Figure 12.4). This technique also allows for post recording changes to the pickup angle. Many of the previous surround sound techniques occupy a considerable amount of space, but the double mid-side technique has the advantage of being quite compact.

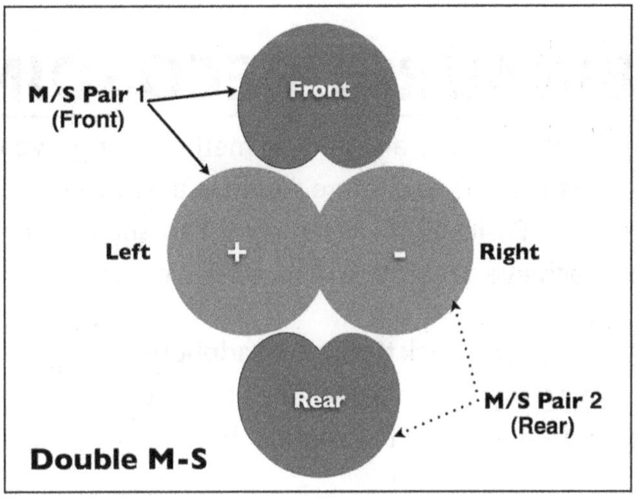

Figure 12.4: Double M-S
© 2023 Bobby Owsinski

As you might think, there are many more channel-based surround miking configurations, but most of them take up a lot of room, are finicky to set up, and can get quite expensive thanks to all the equipment required. Plus, the recorded soundfield is stationary and doesn't move as the head turns, as is required for virtual reality applications. These techniques are falling by the wayside as the need for true immersive audio environments are now a staple of virtual reality and gaming technologies.

Channel-Based Microphones

There's actually only one channel-based microphone available and that has the added twist of added height and rear center channels.

The Holophone was designed and patented by musician/sound designer Mike Godfrey and was developed by Rising Sun Productions of Toronto and Canada's National Research Council.

The microphone (see Figure 12.5) is available in a number of different models and sizes intended for a wide variety of applications. The top-of-the-line H2-PRO features a 7.5 x 5.7-inch fiberglass epoxy ellipsoid that looks something like a giant teardrop.

This 7.1 ellipsoid holds eight omnidirectional microphone elements, five in the now-standard 5.1 multi-channel fashion with the front center element at the tip of the teardrop, plus one at the rear of the teardrop for the center rear channel, one on top for height, and an element internally mounted in the ellipsoid for the LFE channel.

Figure 12.5: The Holophone H2-PRO Courtesy of Holophone Microphone Systems

Because it's just a single piece of equipment it's relatively easy for the sound designer to collect samples in the field; just point the mic, and the Holophone does the rest.

Holophone also offers the 5.1 H3-D, the H4 SuperMINI, and the PortaMic 5.1 for camera-mounted applications. Check out their website at www.holophone.com for more detailed information.

BINAURAL RECORDING

Binaural sound is a recording method using two microphones placed at the approximate position of a person's ears to capture sound in a way that simulates the way that sound waves reach the ears of a listener. Basically, if we can record an audio source the same way that someone listens to it, then we can achieve a realistic 3D immersive effect even from a stereo recording.

When played back through headphones, binaural audio can create the impression that the sounds are coming from specific locations in the environment, allowing listeners to perceive a sense of direction and distance from the sounds they are hearing.

That also happens to be one of the downsides of binaural recording in that the only way to achieve the affect is by using headphones.

Yet another is that binaural doesn't work very well in virtual reality environments, 360° videos or any other interactive 3D media in that users need to turn their head and still have the soundfield locked to the front. In other words, it only really works well if you keep your head still while you're listening. An extension of binaural recording called "Dynamic Binaural Sound" will move as the head does, but it's still in the research stage and not yet commercially available as of the writing of this book.

Binaural Microphones

There are numerous binaural recording systems available, but here are three that cover the different styles available.

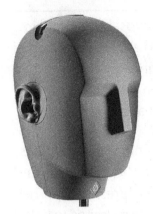

Neumann KU 100 Binaural Dummy Head

The KU 100 simulates the average size, density, and shape of a human head, and even includes accurate reproductions of human ears (see Figure 12.6). With an omnidirectional condenser microphone located in each ear, the unit is capable of recording high SPL levels with very low self-noise. According to Neumann, the playback is also loudspeaker compatible.

Figure 12.6: The Neumann KU 100
Courtesy of Georg Neumann GmbH

3Dio Free Space

3Dio does away with the simulated head and keeps the ears as part of their microphone systems. Several models are available, including the Free Space (see Figure 12.7), FS XLR, and FS Pro II, all with the same basic design but targeted at different applications.

Figure 12.7: 3Dio Free Space Pro II
Courtesy of 3Dio

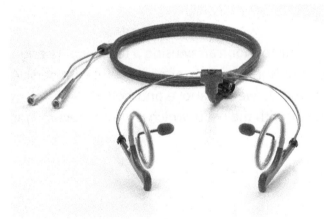

DPA 4560 CORE Binaural Headset

The DPA 4560 takes a different approach by using the user's own head and ears as part of the recording system (see Figure 12.8). While the setup is dead easy, the problem is that the user must stand still while recording and refrain from any personal noises like coughs, teeth grinding, or even heavy breathing.

Figure 12.8: DPA 4560 CORE
Courtesy of DPA Microphones, Inc.

DIY Binaural Recording

As you can see with the DPA 4560, you can get the same effect by mounting a small omnidirectional mike close to each ear. While this will work and is great for experimentation, the big downside is that the playback will be personalized to you only, as each person has a different head shape, size and weight. Also, cheaper condenser microphones that will fit the application are subject to high levels of self-noise.

AMBISONIC RECORDING

Ambisonics is a immersive sound technique that uses a number of microphones closely arranged in a spherical configuration to capture sound from all directions. The captured sound field is then encoded into a special ambisonics format, which can be decoded and reproduced through a set of speakers or headphones, allowing listeners to perceive a sense of direction and distance from the sounds they are hearing. As a bonus, how we perceive the recorded audio signal can be changed after the fact.

This is actually a format that was developed way ahead of its time and is only finding its usefulness now. The science behind the technology was developed in the UK at the University of Oxford and the University of Surrey in the mid-1970s by Michael Gerzon, Peter Felgett and Geoffrey Barton. Their theoretical design was developed into a practical microphone system by Calrec Audio Limited with the first Soundfield mic launched in 1978.

The problem was that from the beginning, Soundfield was a four channel system. Although theoretically perfect for location recording and sound effects, field recorders of the time were mono, and multi-channel portable recorders capable of capturing the mic's four outputs weren't developed until the mid-1990s.

With the advent of virtual reality, 360° videos or other interactive 3D media, along with modern digital technology that streamlined the decoding process, ambisonics has finally found multiple environments where it can shine.

The Microphone Array

As stated before, ambisonic recording is based upon a 4 microphone array using a tetrahedral design (see Figure 12.9). This is what's known as "first-order" ambisonics, and it provides a fully immersive sound experience, capturing sound on a horizontal plane, as well as above and below the listener. Unlike binaural, the sound can also be configured to move with the head, which is why it's favored for virtual reality audio.

Figure 12.9: The SoundField tetrahedral microphone array
Courtesy of Soundfield by Rode

While the 4 microphone/channel array works perfectly well, there was a desire to improve upon the theme. The next development was an 8 microphone array, which is known as "second-order" (which has 9 channels). This offers several significant improvements over first-order ambisonic microphones because it's better at preserving the perceptual cues so that a listener can locate sound sources more precisely. It also provides a much larger sweet spot for listeners.

While first-order ambisonic microphones have a sweet spot about the size of a human head, a second-order microphone is capable of a sweet spot suitable for multiple listeners.

The next step up is a "third-order" array with 16 transducers. The only example of this is the Zylia ZM-1 with 19 calibrated MEMS (Micro Electro-Mechanical System) sensors instead of microphones. MEMS sensors are similar to electret condenser microphones except they're built onto a silicon chip with provisions for a preamp and A to D convertor built in.

A Format

The raw output from an ambisonic mic is what's known as A format and has to be processed into B format before it can be actively used by your DAW. The reason is that the spacing between capsules of different ambisonic microphones isn't standardized, and is therefore specific to each model of microphone and can't be interchanged.

B Format

While stereo defines the world of sound in just two dimensions, B format defines it in three. B format contains information to define front to back (first dimension), left to right (second dimension), and up and down (third dimension). Essentially, B format defines a sphere with four elements: an X plane (front to back), a Y plane (left to right), and a Z plane (up and down), all with a central reference called W. Because all options are based on the same reference information, there are no phase-difference issues to contend with.

Just like A format, B format also uses 4 channels, but it isn't the same thing. B format takes the raw output signals and perfectly aligns them so the spacing in between the capsules is compensated for. That means that the microphone parameters can be manipulated in software later. For instance, you can change the pickup pattern or render the output to anything from mono to 7.1.4 speaker configuration commonly used in Atmos mixing.

A confusing part of B format is that there's actually two standards - Furse-Malham (FuMa) and AmbiX. These are similar but not interchangeable since the track order is different. AmbiX is the newer standard, although most ambisonic plugins today will work with both.

Ambisonic Microphones

As stated before, ambisonic microphones have been around since 1978 and they've come a long way. Let's start with the original.

SoundField MKV Microphone and Model 451 Decoder

While the SoundField microphone has been around since the 1970s in a stereo version, the microphone truly became a surround microphone when the MKV model and the model 451 surround processor were introduced (see Figure 12.10).

Figure 12.10: The SoundField MKV mic with processor
Courtesy of Soundfield by Rode

The SoundField microphone employs the standard four-element tetrahedral array, but it can be electronically controlled from a supplied analog preamp/controller connected by a long multicore cable. The mic is rather small and unobtrusive, considering the number of capsules employed, and can easily be placed in most miking situations, even in the supplied shock mount. SoundField now offers a number of controller units, all of which combine a microphone preamp and the appropriate electronics needed to control the various parameters of the MKV.

The heart of the processor lies within the SoundField controls. It offers some unfamiliar parameters usually not associated with a microphone. For instance, Azimuth allows complete electronic rotation of the microphone, Elevation allows for plus or minus 45° of continuous variation of the vertical alignment, and Dominance is a form of zoom control that gives the effect of the mic either moving closer to or farther away from the sound source. There is also an In/Out control for the SoundField controls, as well as a B-format Input switch for using the controller with prerecorded B-format material.

SoundField now offers a number of software packages, which allows the processing to take place within your DAW via a plug-in. See www.soundfield.com for more complete information.

Soundfield SPS200

The newer Soundfield SPS200 Software Controlled Microphone uses the same SoundField multi-capsule technology as the high-end models in the SoundField range (see Figure 12.11). This relatively lightweight microphone is powered by standard 48v phantom power and incorporates four low noise, studio grade condenser capsules. A short break-out cable outputs the four signals generated by the SPS200 at mic level on four balanced XLRs.

Figure 12.11: The SoundField SPS200
Courtesy of Soundfield by RØDE

The SPS200 is the first SoundField microphone that doesn't require an accompanying control unit. All the processing such takes place in the Surround Zone software which is included with the SPS200.

RØDE NF-SF1

RØDE purchased the SoundField brand in 2016, and the NT-SF1 is the first collaboration between between the companies (see Figure 12.12).

Figure 12.12: The RØDE NF-SF1
Courtesy of RØDE Electronics

Sennheiser AMBEO VR

AMBEO is Sennheiser's product line for immersive 3D audio that covers products and technologies for the entire audio signal chain, from capture to mixing and processing to reproduction. The ABMEO VR microphone (see Figure 12.13) differs from other first-order ambisonic microphones in that it uses Sennheiser's excellent KE 14 condenser microphone capsules.

Figure 12.13: Sennheiser AMBEO VR
Courtesy of Sennheiser

Core Sound TetraMic And OctoMic

Core Sound makes a range of very compact first and second-order ambisonic microphones. It's OctoMic (see Figure 12.14) is one of the few second-order microphones available at a very reasonable price.

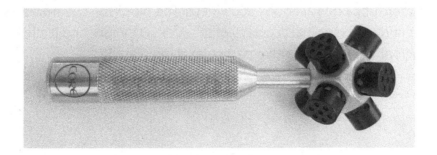

Figure 12.14: Core Sound OctoMic
Courtesy of Core Sound

Zylia ZM-1

The Zylia ZM-1 (see Figure 12.15) is the only available third-order ambisonics microphone featuring 19 MEMS electronic sensors instead of microphones. As a result of this breakthrough technology, the signal is digital from the start and is instantly converted to B format. The output is via a standard USB-A connector that can be connected directly to your computer.

Figure 12.15: Zylia ZM-1
Courtesy of Zylia SP

Zoom H3VR

The Zoom H3VR (see Figure 12.16) is a unique device in that it includes not only the microphone but the recorder as well. A to B format encoding is done on the device and includes a built-in ambisonics decoder for monitoring with headphones.

Figure 12.16: Zoom H3VR
Courtesy of Zoom North America

Ambisonics Software

Because of the problem of non-standardized capsule placement in microphones developed by different manufacturers, each microphone manufacturer also includes a software app or plugin to encode the A format output from its microphone to B format. Most will also allow for input from an external B format stream as well. More and more of these types of plugins are coming on the market, but the following are worth noting.

ABMEO A-B Converter

The AMBEO A-B Converter is the free tool that converts the output of the Sennheiser AMBEO VR Mic into Ambisonics B-Format. After it's converted the signal becomes directly compatible with the spatial audio format of both YouTube and Facebook 360 videos. The plugin also features a special correction filter that's meant to improve the spatial accuracy of the AMBEO VR mic, and it allows you to adjust the orientation of the microphone to align with a particular camera angle.

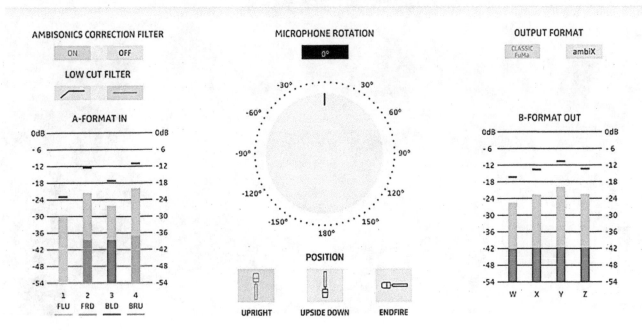

Figure 12.17: AMBEO A-B Format Converter
Courtesy of Sennheiser North America

SoundField by RØDE Plugin

This free SoundField by RØDE plugin replaces Soundfield's original SurroundZone2 and offers a new approach to ambisonic processing. Instead of using the various digital correction filters used in previous processors, it uses frequency-domain processing to deliver improved spatial accuracy at all frequencies. Like most other plugins of this type, the user is able to set up mixes for all the common surround-sound formats from 5.1 through to 7.1.4 and even custom configurations.

dearVR AMBI MICRO

AMBI MICRO is a free plugin released in cooperation with Sennheiser that has been optimized for the AMBEO VR microphone. The plugin converts ambisonics A-format to B-format and binaural audio while also featuring head tracking with a VR headset, and contains the ability to work with DearVR, Facebook360, YouTube360, and the Neumann KU100.

Harpex-X

Harpex-X is different in that it wasn't made to work with a specific microphone and it's intended for film postproduction. Since it's able to produce surround decodings with far greater channel separation than other decoding plugins there's a larger playback sweet spot, which is prefect for large venues like movie theaters and fixed installations running Dolby Atmos, IMAX and Auro 3D.

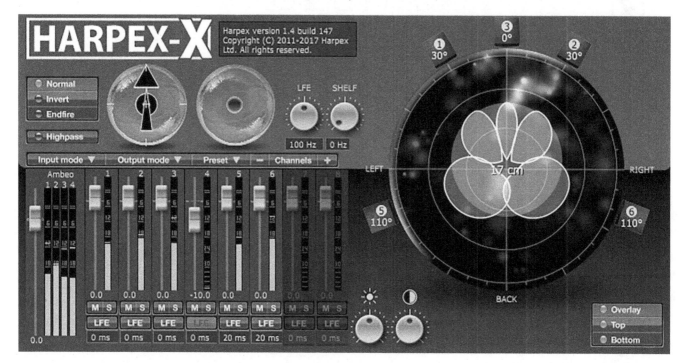

Figure 12.18: Harpex-X
Courtesy of Harpex Audio GmbH

PART II

THE INTERVIEWS

13

CHUCK AINLAY

Chuck Ainlay is one of the new breed of Nashville engineers that brings a Rock and Roll approach to Country music sensibility. With credits like George Strait, Dixie Chicks, Vince Gill, Trisha Yearwood, Patty Loveless, Waylon Jennings, Wynonna and even such Rock icons like Dire Straits and Mark Knopfler, Chuck's work is heard world-wide.

Do you have a standard setup that you use when you track?

To some degree. I have favorites that I start with and if that doesn't work then I try other things, but it depends a lot on the type of music that I'm trying to do. If it's a Country thing or if the song is kind of a '70s thing or something like that, I might try different miking techniques to capture that type of sound.

From a Nashville perspective, and maybe where I should start, most things are pretty typical. Somewhat of a departure is that I use two mics on the bass drum and the snare drum. I don't think that's radically different from what other people do but it's not typically what everybody here does. Usually I'll put an AKG D-112 inside the bass drum slightly off-center from the beater head and back about 6 to 8 inches and generally pointing towards where the beater strikes the head. It's not straight in the middle; it's usually off-center a little bit. Outside of the bass drum I'll place a FET 47. Usually drummers have a hole cut in their front head, and I prefer that rather than no front head at all. It gives you a bit of that almost double-headed bass drum sound. I'll put the outside mic off-center once again, away from where the hole would be cut, then it's just a matter of time spent dampening the drum with some soft materials to try to get however much deadness you want out of the drum.

Usually then I'll put either some mic stands or chairs or something that I can drape some double thick packing blankets so that it makes sort of a tunnel around the bass drum and helps seal off some of the leakage into that outside microphone.

How far away is the 47FET?

That varies a lot, but usually about 8 inches. Both of those mics go through Neve 1081 modules. I have a rack of those and a fair number of other preamps that travel with me when I'm tracking because I typically avoid using the console preamps for many of the important things and just go straight to tape or hard disc from the outboard modules. The console is used just for monitoring.

I don't usually use any kind of compression on the bass drum, or any of the drums actually, with the exception of the ambient mics. I don't use gates on anything except for the toms.

What do you do for the snare?

The snare drum usually gets a 57 on the top with usually a 452 with a 20dB pad on the bottom, although that mic varies to some degree because there might not be a 452 with a 20dB pad in the studio where I'm tracking. The top mic will usually go through a 1081 again but the bottom mic I'll bring up on a console mic pre. It's always flipped out of phase and combined with the top mic so that I just have one snare drum track. I like to commit to that because the two mics together really makes the snare drum sound and I don't like to leave that open for judgement later.

Do you do the same thing with the kick?

No, the kick I'll usually leave on separate tracks because that, to me, is one of those things that I can refine later. If I have a lot of leakage on the outside bass drum mic I can spend a bit of time gating that out or cleaning it up on the DAW if need be. That's one of those things that when you start getting into the nuts and bolts of your bottom end with the bass, having those two mics separate will allow you to change how much attack you have in the bass drum or how much "oomp" there is.

On the snare drum I like to put it to where the rear of the mic is rejecting the hi-hat as much as possible but it isn't in the way of the drummer. The main thing with miking drums is for the drummer to never think about hitting a mic while he's playing. The mic usually comes in somewhere between the hi tom and the hi-hat, but I like to somehow get the rear of the mic towards the hat for the most rejection. It usually is pointing down at sort of a 45 degree angle. I find that the more I angle it across the drum the better side-stick sound I'm going to get. If it's pointed down too straight at the drum then the side-stick becomes too much of a high frequency click rather than a nice woody sound, so if there's a lot of side-stick, then I might have to position the microphone more for that instead of rejection of the high-hat. Once again all this stuff varies from session to session.

High-hat I vary between a 452 and a KM 184. It depends on what kind of sound I want. If I want a chunky sounding hat the mic will usually be over the hat pointing out across the hat somewhat away from the snare drum, so if you're the drummer it would be on the other side from where you're hitting it. If I want an airier sound, I'll move the mic more and more off to the side of the hat to where it's not even over the hat to get that paper thin sort of sound. The only thing that you have to be aware of is the wind blast that might happen when he pumps it.

For toms, the microphone choice there varies the most of anything. It will vary from a 57 or a 421, although I've been using these Audio Technica ATM-25's a lot lately. Sometimes if I want a beefier, warm sound I'll go to a condenser microphone, which can go from a 414 to an 87 if I want a sort of fat 70's sound, to Sony C-37's if they're available and working (laughs). Once again Audio Technica makes a clip-on condenser, the 8532, that I've had a lot of success with. It has a lot of isolation and doesn't have that huge proximity effect that you get from a lot of other condenser mics. It also works great on acoustic guitars. Between that and an AT 4033 I get an amazing acoustic sound.

Also on toms, I always put gates on the inserts of the tom channels. What I do is use these little contact mics that were intended to be trigger microphones for triggering sound modules for drums and plug them into the key side of the gates for the toms. Whether or not I turn on the insert depends on whether I want the leakage on the toms or not. Toms add so much to the warmness of your snare drum and bass drum but this way I have a really solid trigger on the gates and I don't miss the nuance-type fills. I don't necessarily always use it, but it works so good when I need it. Usually when I do gate toms it will only be 6 to 10dB of reduction. I don't gate them to nothing. I usually use the console mic amps so I can do this.

I normally place the mic between lugs of the tom. If you get over one of the tuning lugs you get too much of the flap from the drums. Drummers usually don't tune their toms perfectly so they don't ring on forever. They'll intentionally detune them slightly so they sort of bend away and stop ringing quicker, so if you split the lugs it sounds better.

Also I try to not get too close to the head. You're compromising between leakage and tone, but if you get too close you're just going to get attack and no warmth out of the tom. It's usually somewhere between 2 1/2 to 4 inches; probably closer to 3 inches. If I take my 3 fingers and put it between the mic and the head, that's usually a good starting place. Sometimes I'll mic underneath as well but that's rare.

Do you flip the phase on the bottom mic when you do that?
Yeah. If we're really going for like disco drums then we'll take the bottom heads off and mic it, but that's a pretty old sound.

What do you use for overheads?
That varies a lot but it's either 414's, or the stereo Royer SF-12. Ever since I started using that thing my drums have sounded so much more real. I can sort of rely on that for the drum sound and then fill it in with the close mics rather than the other way around.

I was just going to ask how you approached setting up the kit balance.
Well it depends on whether I'm using a stereo mic overhead like that, which then you can use it for the main kit sound. If you're using spaced pairs where you're just miking cymbals, then it doesn't work at all. It depends on the intent and if you want this really in-your-face closed mic thing or if you want the drums to be more set back and more real sounding. If you're going for that '70s/'80s tighter sound, then you'd put the mics over the cymbal. If you want them more real sounding then you'd go for the stereo mic.

How about room mics?
I usually put them about 4 feet out in front of the snare drum. Not the kick drum but the snare drum. I'm sort of splitting the positioning between the hat and the bass drum with what I call a mid-field mic. I'll usually put up one of the mono Royer mics (R-121) and use some severe limiting on it with an 1176. That just sort of brings in the drums as an overall picture and it really adds a lot of meat to them.

Then about 12 feet away (sometimes less) in front of the drums I'll put up a pair of 149's or a pair of these Joe Meek microphones; the JM47's. Sometimes I'll do like a Led Zeppelin thing with one mic picking up the high tom, snare and hat and the other one picking up the floor tom and ride cymbal and the JM47's sound great for that.

How much time do you usually take on getting sounds?
I'm pretty fast. Maybe an hour at most and that's for getting things situated in the room. After that we might change the drums before each song and it'll be 15 minutes. They might change out entire kits, but usually the EQ just works and I might tweak things a bit as they're running the song down. It's more about changing stuff out rather than tweaking things on my side. I certainly don't take 2 days like they did on the old Fleetwood Mac albums.

Are you EQing while you're tracking?
Actually, I EQ the drums a lot while I'm tracking but I do avoid compression. I try to do minimal EQ on most things but bass drum has always been one of those things that usually ends up taking a lot of EQ.

How about bass? Are you taking it direct with an amplifier as well?
In Nashville most of the guys have their own rigs because it's so session oriented. They'll come in with these amazing racks full of great gear. They'll have a Telefunken mic pre and an LA-2A or Tubetech, which is sort of typical, so usually all I have to do is take a direct line from them to the recorder and I don't mess with it in between.

Sometimes if it's just not happening for me, I'll say "Hey, I have this really great direct box" and I'll run it out to him. It's an Agular and I love it on bass. It just sounds so big and real. I don't carry a bass amp, but I really like the old Ampeg B-15's because you can distort those things if you want. Big rigs don't work for me, but the little guys do. Usually I'll just put a FET 47 or the Neumann 147 in front of it.

If at all possible, I really like the sound of the bleed in the room. If I have a great bass player that I know I'm not going to move a lot of notes (which is most of the guys in town), I'll let them have the amp right next to them. The room mics for the drums pick it up and you get this big bass sound that fills up the whole stereo image instead of something that's just right in the middle.

I was just going to ask you about leakage.
I like it. Once again though, if you're dealing with a band where you know you're going to be moving notes, then you have to isolate it. Some places have rooms where you can open it up enough to where you can put room miking on it and it's really nice to get that spread on the bass.

Is the approach to recording different in Nashville from other recording centers like New York or LA?
I don't think they differ that much except for the fact that we do so much session musician stuff. Typically we play back a song demo at the beginning of the session and it dictates a whole lot about how the recording is going to go down. They may play it down and then change the form of the song afterwards, but in a lot of ways the demos are somewhat copied for the master. Many times the musicians will play it down the first time and that will be the take.

We're not just talking about a small section either. We're talking about bass, drums, two guitars (one may be acoustic), fiddle, steel, two keyboards (piano and organ) and vocal. This all goes down live.

Is this typical?
Yeah. You have to be ready to get the first take because they'll have it ready by at least the third, so when you ask how long it takes to get drum sounds, it's got to be fast. It's a blast to cut tracks in Nashville because you're so on fire. You can't make a lot of changes as things are going down. You've just got to make a mental note in between takes. If you've got to move a mic, you've got to do it as the musicians are listening back to a take. You're really flying around. It's a blast.

That's so different as compared to the normal rock way of tracking.
Where they take all day to get a track and it's just bass and drums and guitar and then you strip it down to the drums and replace the bass and guitar? That's drudgery. In Nashville tracking is one of the most enjoyable things you can possibly ever do. Not only do you have a bunch of really great people that you're hanging out with, but some of the most talented musicians in the world too.

A typical studio will have at minimum a piano room, a room to isolate an acoustic guitar (sometimes you might jam two acoustic guitars and a fiddle in that same room), and a room for the vocalist. A lot of the tracking rooms here are built that way because often the guys go for their solos as the track's going down, so they need some degree of isolation if they want to fix a bar or two of their solo later.

That must mean that you don't spend much time doing overdubs.
Heck no. Most of the track is done when you finish tracking. The singer might sing the song three or four more times at the end of tracking and go home, then we just comp vocals, do some background vocals and maybe there might be another guitar added and maybe strings or horns, but usually we just go straight to mix. Nowadays we spend more time tuning the vocals than doing overdubs (laughs). If we only had singers (laughs some more).

How do you approach recording vocals?
I still base everything around the vocal. To me you have to find the microphone that fits the vocalist the best because if you get a great vocal sound you're going to bring up everything to match that. If the vocal is so much bigger than everything else then you are going to work on everything else until it's as good as the vocal. If the vocal sound sucks then nothing is going to sound good because you don't want to overpower your vocal with the band.

My favorite vocal microphone is a 251, a C-12 works about as well as a 251, although I like the 251 better. A real U 47 (not a Nuvistor version) sometimes works and so does a FET 47 too, and a U 67 is always a favorite.

Do you approach steel any different than electric guitar?
Once again I'm fortunate to have the best steel players in the world available. There are no others that even come close. With Paul Franklin, who's probably the most in demand steel player, a 421 works perfectly, but I'll always place the mic off-center from the voice coil. He and I have just figured that out

over the years, although he has a whole rig of stuff that he tweaks until he gets what he wants and it just works wonderfully.

For electric guitar it's usually a 57 off-axis pointed in (toward the voice coil) and a 67 out from the speaker about 6 to 12 inches. A lot of the guys have big rigs and usually just a couple of 57's off-axis will cover it because they use so many effects to get a stereo spread that an ambient mic isn't worth-while.

What's your approach to acoustic guitar?
In Nashville we tend to make pretty sounding acoustic recordings. I guess I'll use an AT 4033 or a 452 if I want that bright Nashville sound. If I want more of a richer sound I'll use either a KM 84 or KM 56 or one of the new 184's. In all instances the mic is pointed at where the neck joins the body and then out about 5 or so inches. I usually use a second microphone that moves around a lot. It's usually a large diaphragm mic that's placed away from the guitar. That varies so much. A 67 is probably my preferred mic for that, but an Audio Technica 4033 or 4050 works well too.

I'll start out straight in front of the sound hole. If that's too boomy I'll either move towards the bridge or lower or sometimes above the soundhole above the cutout. Sometimes off the shoulder near the right ear of the player works. I might just put on a pair of headphones and move the mic until it sounds great. That's about the only way that you can mic an acoustic guitar; you just have to listen.

I'm always trying new things. This stuff will be valid today but I may be doing something different tomorrow.

14 MICHAEL BISHOP

There were few more versatile engineers than the late Michael Bishop, easily switching between the classical, jazz, and pop worlds with ease. Shunning the use of massive overdubbing, Michael instead mostly utilized the "old school" method of mixing live on the fly with spectacular results. A former chief engineer for the audiophile Telarc label, Michael's highly regarded recordings have become reference points for the well done and have resulted in 8 Grammy awards.

I know you do a lot of sessions where the mix is done on the fly either direct to 2 track or multitrack with no overdubs. How do you handle leakage?
I let leakage be my friend. Leakage is inevitable for the kind of recording that I'm doing because I like to keep the musicians together as a group in the studio rather than spreading them all out with isolation for everybody. That means that there's plenty of leakage and I just deal with it. I don't have to have the isolation because typically I'm not doing overdubs and replacement of tracks. We fix things by doing new takes to cover the spots that we need to cover. They'll take a running start at it to cover the measures that they need and we'll edit it later, which is very much a classical orchestra style of recording.

By keeping people close together the leakage generally becomes less of a problem. The further apart that you get the musicians and the more things that you put in between causes delays and coloration, particularly on the off-axis side of a microphone, which is already colored. This only exaggerates the effect of the leakage. That's something I learned from John Eargle's very first microphone handbook. Using omni's, I learned how to work with the leakage in the room and make it a pleasant experience instead of something to be avoided.

Are you using primarily omni's?
I like to start with an omni before anything. Now there are particular instances where I'll immediately go to something like a figure 8, but I'll use figure 8's and omni's more than anything.

Does the type of music you're recording determine your microphone selection?
Of course, because there are certain things that the musicians or the producer or even the end listener expects to hear on a particular style. Like if it's a straight ahead blues recording, then there's a sort of

sound that's typical of a drum kit on that kind of recording, so you use something fairly raunchy like a 414 in places, where on a jazz date I might use a 4006 or a Sennheiser condenser. Then there's the plain old thing of putting a 57 on a guitar amp where it just works, so why reinvent the wheel?

What's the hardest thing for you to record?

A very small acoustic ensemble or a solo acoustic instrument, but particularly small acoustic ensembles like a string quartet. They have less to hide behind and I have less to hide behind (laughs). Actually, I think recording a symphony orchestra is fairly easy in comparison to a string quartet. It's pretty easy to present this huge instrument which is an orchestra because just the size and numbers can give a good impression almost no matter what you might do. You have to really screw it up to do badly there. But a string quartet is really difficult because you can hear every little detail and the imaging is critical, particularly if you're working in stereo. It's really hard to convey a quartet across two channels and get proper placement and imaging of that group. That's one of most difficult tasks right there. It becomes easier in surround.

Is your approach different if you know that the end product will be in surround instead of stereo?

Absolutely, because a stereo recording has to present width, depth and all of the correct proportions of direct to ambient sound, and in surround you have more channels to present those aspects.

Is your approach similar to the norm when recording an orchestra, with a Decca Tree and house mics?

My approach on an orchestra has never been with a Decca Tree. It started out very much following along in the steps of Jack Renner, who originally hired me at Telarc, and who developed the well-established Telarc sound on an orchestra that the label is known for, so I needed to be able to continue that tradition of the so-called "Telarc Sound." At that time when I first started, Jack was typically recording with 3 omni's across the front of an orchestra and perhaps two omni's out in the hall and that was it. I followed along in that tradition until I came up with something of my own to contribute.

I changed it from the 3 omni across the front to 4 omni's across the front with the two center mics being 24 inches apart, so it was a little like a half of a Decca Tree in the middle, but the positioning was very different. A Decca Tree typically has that center front M50 (or whatever microphone) well up above the conductor and into the orchestra somewhat. That, to me, presents a sort of a smear when the mics are combined because of the time-delay differences between the front microphone that's ahead of the other two mics. These delays destroy some of the imaging and produce a bit of comb filtering to my ears, which is why I never liked the Decca Tree. If you were taking those microphones and just feeding three separate channels it would be OK, but that's not how it's used.

So having the microphones in a straight line across the front gave a clearer sound and I could get perfectly good focusing with careful placement of those two center microphones to get good imaging through the middle of the orchestra. That's one thing that I always look for; the imaging across the orchestra that lets me feel where each musician is on stage. Use of spot mics pretty much destroys that, so I tend to shy away from using them.

Anyway, I quickly moved from that to using a Neumann KU-100 dummy head in the middle as part of the quest for better imaging across the middle. It got in there by accident. I was really just trying it as sort of a surround pickup and experimenting with binaural and one time I got brave and threw it up there in front over the conductor and that became the main stereo pickup on the orchestra with omni outriggers out on the flanks.

What do you do for the hall?

For stereo I continue to use a couple of omni's out in the middle point of the hall, but when I started actively doing surround some years ago, that wasn't satisfactory any more for the rear channels. They were too far removed in that they got the reverb but the sound was always somewhat disconnected from the front channels.

So early on I brought my surround mics fairly close up to the stage and started to experiment with a number of different setups which I'm still fine-tuning, and I probably forever will because it's such a difficult thing to capture properly. Often those surround mics are anywhere from 15 to 20 feet out from the edge of the stage depending upon the house. They're not out very far at all. The most common surround pickup that I've been using is two M/S pairs out there, looking forward and back on each side. They would be assigned to Left Front/Left Rear and Right Front/Right Rear as far as the decoding output of the double M/S pair, so I'll be using a figure 8 and a supercardioid, usually the Sennheiser MKH 30 and MKH 50, which are the easiest ones to use in this case. Often I'll be using the Sennheiser omni's as the flanks to the KU-100, although sometimes they'll be Schoeps depending upon the music and the hall.

How much time do you have to experiment on sounds in a new hall?

Luckily I'm pretty good at enrolling people to go along with my crazy ideas. I'll get the orchestra management to allow me to hang microphones during the orchestra rehearsals in the days leading up to the session. I'd like to have a good day during their rehearsal time to experiment with placement. I'll always get up on a ladder and get up in the air to listen for where the sweet spots are. There's that magic blend up there that just doesn't seem to happen out in the house, so I'll find the right height and distance for my mikes relative to the orchestra and try them there during the rehearsal if at all possible. This is probably against all AF of M rules, but I don't ever record when doing that sort of thing so there's no danger of using material that isn't paid for. This is all due to the good graces of the management of the orchestra that I can even attempt this.

In cases where we can't do this, all I have is what the AF of M allows, which is technically 5 minutes at the top of the session. That's one of the drawbacks of recording in the States, which is where I work most of the time. Overseas you have the luxury of being able to have the mics up and do extensive sound checks during rehearsal. Of course, the time there is not as tight either because you're not restricted to the 3 and 4 hour typical orchestral session.

Have you done any experimenting with the surround mics that are presently available like the Soundfield?

I've used the Soundfield on a number of sound effects recordings. I've tried it briefly on a couple of sessions and came to the conclusion that I really didn't like that much of a point source for picking up either stereo or surround. While it was technically correct and it's a wonderful way of manipulating the sound after the fact, there's something about the musicality of it that I didn't like. It doesn't have the width that I look for either, which is something that I'm accustomed to getting with spaced omni's and the various combinations that I use.

The other thing that I've tried is the Schoeps Sphere, which is an excellent means of recording surround particularly in the Jerry Bruck combination of figure 8's combined with the brightening center in the sphere, but it's somewhat limited for the type of recording that I typically do with an orchestra.

Do you start with the same setup every time?

Every session is unique, but there are places that I visit regularly [in an orchestral setting] so I know where to start on those. There are still a lot of things that need to be different given the piece of music that we're recording.

How about the electronics?

The electronics are steady. I use a standard setup of Millennia Media preamps all around.

What's your approach to doing an ensemble in the studio?

There the performer is taken out of the natural setting of a performance space, which you are trying to recreate because you don't want to represent a studio sound usually. Since people are screened off and set up more for sight lines than for anything else, you're not necessarily presented with a nice acoustic blend out in the studio, so there isn't a whole lot to record ambience on. If I'm in a situation where I'm in a tracking and overdub situation, I will often record at least a three track pickup. Like for a sax overdub, there'll be a single pickup for the sax with at least a stereo ambient pickup, which will give me something to work with latter on.

How far away is the ambient mic?

Oh, not very far away at all – maybe 6 feet. If that doesn't get enough ambience, I might change the mics to cardioid and flip them around to face away from the instrument. One thing that I've been working with a lot has been double dummy heads (Neumann KU-100's); one facing the instrument or ensemble and the other with its back to the ensemble and pointed upward and away from the group up into the room.

I tried this a couple of years ago with a small acoustic ensemble in a little performance space outside of Baltimore. I had a second borrowed head that I just put out there to try. It didn't sound right facing the group, but as soon as I turned it around facing the room (and this is with a spacing of only 3 to 5 feet between the two heads,) it became a 3D sort of experience with only 4 channels. I did add a center mic (an MKH 50) to solidify some stuff that was closer to the stage and that helped.

Where the mics placed back to back?
They were back to back with hardly any spacing. If you listen to the rear channels only, it sounds like almost the same recording as the front channels except that the timbre has changed because it's coming in at the back of the head. The high frequencies are somewhat muted and, of course, the delays are somewhat different. The combination with the front head was just about ghostly.

15

J.J. BLAIR

Grammy-winner J.J. Blair has worked with a variety of artists that include Rod Stewart, Johnny Cash, June Carter Cash, Jeff Beck, George Benson, P. Diddy, Smokey Robinson, and many more. I've worked with a lot of great engineers, but only J.J. has that combination of musicianship (he's an excellent guitarist), engineering skill, audio electronics knowledge, and attention to detail that makes him totally unique.

You record a lot of different styles of music. Is your approach different for each style?
It depends. I have things that I like to do but the style of the music is always my first consideration. Am I the producer or am I working for another producer? What does that producer have in mind? What does the artist have in mind?

If it's not up to me, then I'm trying to get an understanding of who I'm collaborating with to get an idea of what I'm going for. I find it very helpful for them to play me specific references rather than giving me esoteric terms that can mean a number of things.

Like "I want it to sound big?" Does that mean Led Zeppelin big or something else big? Give me a concrete example. So that affects my approach.

My approach is also going to be affected by where I'm recording and what gear do I have available. Sometimes I'm not in Los Angeles and wherever I am I can only bring a finite amount of my own gear. But ultimately the biggest dictator is the style of music and what suits that type of material.

Let's go there for a second. If you were going to travel somewhere and you were going to bring gear, what would you bring? Would it be mics and preamps mostly?
I tend to bring a couple of my personal mics for vocals based upon what I think will suit that particular vocalist, and I will bring my vocal chain. I always travel with a [AKG] D12 for the bass drum unless I really trust that they have a D12 that's in working condition. I always bring a couple of [AKG] D19s - one for snare because that's a thing that I love, and then if there's a piano I really love it as a mono piano mic. I bring a couple of [Beyer] M160s with me because they give me a room sound that I can't argue with.

Again it depends on what they have, but those are the things that are essential for me. I'll also bring a couple of [Sennheiser] 409s with me because that's my first choice on a bass amp and it's my first or second choice on a guitar amp depending upon what we're going for.

Because the vocal is so important and because they're telling the story of the song, I want to have the right mic for them, the right pre and the right compressor that I know will give me not only the recording that I want, but will help them hear themselves in a way that makes them comfortable and sound like a record in the headphones, which I find brings out better vocal performance. When someone is on the right vocal chain, I find that they're much more comfortable singing.

Let's talk about your vocal chain. I know you may change it per the style, but generally, what do you use?

99.9% of the time my signal path is my Inward Connections Vac Rac pre and Vac Rac compressor. There's just unlimited headroom and clarity. I like the ones without the transformer because I want the air on top. It brings out the texture in way that the attack and release functions. All compressors bring out a harmonic element when you dig in. This one does it in a way that colors it the least so that I can hit it hard when I'm tracking and not have to worry too much about coloration.

For the mics, it usually comes down to, is it a U47, a M 49, a 251 or my 367, which is a U67 with an AC701 tube that has more top end than a normal 67. I've personally worked on all my mics and tailored them for vocals. Honestly, I can rarely trust the condition of the mics I run into in most studios.

Let's talk about drum miking for a second. You mention a D12 on the kick and a D19 on the snare. What else do you use?

I like either 251's or C12's on overheads, unless I'm working for a producer who requests something specific. I'm an isosceles triangle guy. I'm not trying to capture the cymbals, I'm capturing the kit, and any close mics that I use are just augmenting the picture that I get from the overheads.

That means I also have to be aware of how much and what type of room sound I'm letting in. I like a larger room so I don't have to worry about a reflection from overhead, but sometimes I'll stick baffles on the back and maybe on the sides just to get an appropriate amount of space around the kit.

It's funny but the longer I've been doing this the more I can hear the ambient sound of the drums on records. Often I think, "That's not really appropriate for this song," so it's something that I've become very sensitive to now. I listen to some of my earlier recordings where I think, "I'm hearing the room but it's not the right room for this material."

How do you handle leakage? There are some engineers that embrace it, and some that try to get rid of it at all costs. Where do you come down on that?

If it's just drum leakage, then I embrace it. I don't like having to mute the toms because I think the resonance is part of the sound of the kit. I never put a ride mic up because the way I do things, you don't need it. Also, because whatever mic I use on the toms will have enough leakage, and I'm going to pick a mic where the off-axis leakage is musical and is part of the overall drum sound.

Sometimes you need things to be so tight and so dry, but that's rarely the case. And you're probably going to be dampening the crap out of the toms anyway in that situation. But yeah, I like a live sounding kit.

On the toms, I tend to use [AKG] 414 EBs. I'll use 87s if I'm in a studio and that's what they have. I love C37As. If I want a lot of off-axis rejection I'll use Beyer M88s. Those are wonderful. And sometimes I'll just use D19s if I feel like a Beatles type of sound.

How much do you compress while you're recording?
It really depends. I'm compressing a vocal to where it sounds right. I like things to sound like a record on the way in [to the DAW]. I want people to be playing inside the record because that's going to change the way they play and the choices they make.

There are certain rooms that have a sound where I'll be absolutely nuking that compressor, but on a vocal I compress it to where it sounds right. I use my ears for that.

I guess that also means that you're not afraid to EQ when recording.
No, I tend not to EQ a vocal though. I rely on the mic for EQ. In the comment section of Pro Tools I will mention the mic, what pattern it is (I'll use the pattern as an EQ - if I want it darker I'll use a Figure 8), and then the settings in case we have to pick up a vocal later. Sometimes an EQ choice on one day won't work on another because there's something about that person's voice that's changed. The more I go from analog to digital, the more I lean on my analog components on the front end to make it sound like a record.

Let's talk about recording electric guitar. Since you're a great guitar player yourself, that changes your approach from other engineers because I would think that you would hear it differently.
I think everybody hears the way a guitar should sound differently. Again, it depends who I'm working with and what their expectation is. I have to get on the same page as far as guitar tone before I ever put a mic on it. I've spent a lot of time doing a lot of tweaking on guitars and amplifiers, so I know when something isn't really cutting it.

I don't really have an issue working with experienced guys. I'm not really asked to work with younger bands anymore. I would get a lot of, "My XYZ into my XYZ really sounds great!" and I'd go, "Try this," and their eyes will light up.

You said you're using 409s on guitar amps, which isn't really a popular choice.
Well, they used to be. I'll tell you why I like 409s. They give you the presence at a slightly higher frequency than a 57. The 57 tends to put things towards 1kHz and the 409 brings it up between 2 and 3k, and then it brings in that low end that people always have to use an [Royer] R121 to get. Now you can do it with just one mic.

Given my druthers I'd put a U67 about 10 or 12 inches off the cabinet. Then it sounds like that guitar amp. That's the sound of so many records that I grew up listening to.

When you go into a new studio, how do you determine the best place to put everything?
I ask the assistant (laughs). I can rely on someone else's experience there. I don't see any point in reinventing the wheel.

What's the most difficult thing for you to record?
I can't think of anything where I go, "That's hard to record." The most difficult thing to record is when there's a communication problem from what somebody wants and what you're trying to do. If somebody gives me a clear idea of the sound they want, I've been pretty fortunate to be able to give them that the majority of the time.

You get to a certain level of competency, and then what you're hearing is opinion. When you're hearing my drum sound, it's not whether that I'm a competent engineer. It's my opinion on how drums should sound and you either agree with my opinion or you disagree with it. If someone lacks competency, you can hear that clearly, but in most cases the engineer has skills. It's just a different idea on how it should sound.

It's like Eddie Van Halen and Jimmy Page. Eddie had an idea of how guitar should be played, and Jimmy had a different idea of how it should be played. Neither is wrong, and there's room for both things in this world.

Do you have a philosophy on recording?
The first rule as an engineer for me is "Know Thy Gear." You have to know what's going to make the correct sound for the application. You try it, and then it either works or it doesn't, and then you can move on to the next thing.

I think shootouts are the death of inspiration. I have friends that have systems set up that you can go, "Let's try this. Let's try this. Let's try this." and you can just push a button and hear it. But I think you really have to intuit, "What's the correct vibe for this." Not, "Let's try five things and make a decision."

There was an engineer in the 90s and 00s that would shoot out 10 pieces of gear to find the best one. I knew people who worked with the guy who spent two days just choosing the kick drum mic. That's not making records to me. The inspiration is gone. The most important thing is the performance and not the sound. That's the main thing that affects my approach, if that makes any sense.

If you had only one microphone to use, what would it be?
I would say and M49. A 67 does everything well, but an M49 does everything well plus it gives me a little more sexiness that I want on a vocal. I always return to the fact that it's always about the vocal.

The Recording Engineer's Handbook - 5th edition

BRUCE BOTNICK

Few engineers have the perspective on recording that Bruce Botnick has. After starting his career in the thick of the L.A. rock scene recording hits for The Doors, Beach Boys, Buffalo Springfield, The Turtles and Marvin Gay, Bruce became one of the most in-demand movie soundtrack recordists and mixers, with a long list of blockbuster credits. Always on the cutting edge of technology, Bruce has elevated the art of orchestral recording to new heights.

How is an orchestral session different from a rock session?
They're mutually exclusive. On a rock date there's more close miking than in orchestral recording, which uses mostly distance microphones. Back when I started in the early 60s, I learned from Ted Keep, Val Valintine, John Paladino, and Armin Steiner where to put the microphones. I learned by watching and listening. They placed a microphone somewhere and I thought, "I'll try that." Also, on the back of almost every record album in those days was a list of microphones used and all that sort of technical stuff. If you heard something you liked you'd go, "I'm going to use a U 47 on the trumpets," so I started trying things to see what would work for me.

At Sunset Sound where I started, you really had to get in close in order to get separation because the room was so small. It was common to put mikes up close to the drums and guitar amplifiers, but at the same time you had to back away when doing strings. A lot of the same things work for me in the studio today. For instance, I went from distance miking over the vibes and tympani to in close to get separation to back out again for orchestral recording, so it's like I made a full circle.

Leakage is something that you're not concerned with then?
No, I'm really not. I like leakage. If it's a good sounding room, leakage is your friend. It's what makes it sound bigger. Let's say I've got 12 woodwinds and I'm using four microphones. In other words one for the flutes, one for oboes, one for the bassoons and one for the clarinets. They're going to be pretty tight, meaning the mic is placed about 5 or 6 feet over them. That's not rock n' roll tight; that's orchestral tight. If you open up that microphone you're going to pretty much hear what that mic is pointing at, and the leakage from the other microphones on the woodwinds make the size bigger on the instrument. Same thing with the overall microphones. If you listen to your overalls and then open up your sweeteners into it, you can control the amount of presence that you want from that distant pickup.

Are you concerned about cancellation from all the open mics?
If they're pointed in slightly the wrong direction from one another you will get cancellation. It's like when you multi-mic drums, you get lots of phase shift between toms. I remember there was an English engineer who had just done the Bee Gee's first album and he showed me what they were doing over there, which was what a lot of people do today. Overheads on the cymbals, individual mics on the toms and getting really tight in and building a drum sound from scratch, rather than being a little distant and getting an overall picture of the drums and then adding things into it. At that time I noticed major phase shift, where by moving the microphones even an inch say from the two center toms, I could change the total character of what was happening. I was always amazed how much things would change especially if you changed microphones or patterns. I used to try a figure 8 next to a cardioid to try to avoid the phase shift.

How did you transition from doing mostly music to doing film work?
Doing movie dates was just something that happened. I did a lot of movies for Disney when I was working at Sunset because Tutti Cameratta (who owned Sunset Sound) was running Disney Records too, so he used to do some pictures there as well. I did all the Beach Party movies and some movies with [producer/arranger] Jack Nitsche, but they were all basically rock n'roll songs for films.

Later on when I went to Columbia Records as a producer, one of the gigs that they gave me was to be executive producer for the soundtrack of *Star Trek – The Motion Picture*, so I was on the stage every day and somehow or another developed a relationship with Jerry Goldsmith, and then later John Williams, and it just started to expand from there. I also got tired of being a psychiatrist for my artists, with albums that should have taken six weeks to make taking over a year while we were forgetting why were there in the first place. It became drudgery and I got burned out, then the movie thing happened and I found it very enjoyable, not to say that it isn't complicated though.

Yeah, but you must get the same satisfaction given that it's relatively quick?
Yes, it goes back to my days of doing things live. That's the way I learned. We always did things live to mono and live to stereo and ran a three track as a backup. Today it's not very different. We go for live mixes on the sessions, so it's basically what I did when I started.

Do you use a Decca Tree?
Sometimes I use it and sometimes I don't. Sometimes I just use three Mathews stands and sometimes I use outriggers. It depends on the score that determines what I need. I don't do everything the same way twice because I find that rooms will change acoustically, depending upon the temperature and humidity, and all of a sudden you have to change the mics in order to compensate for wherever the room's going.

Is this during the course of a session?
Yes. I've had it where the first two days sounded amazing and all of a sudden the third day is as dry as can be and you have to either raise the mikes or go in closer or change them. It's not uncommon, but

it can be shocking when the orchestra hits that first note and things are suddenly different. You have to move really fast.

That must mean that your approach is different every time then.
Yes, I basically have a standard way that I work, but sometimes I will change the microphones that I have on my strings or woodwinds. Basically my percussion mikes and overall mikes stay the same from show to show.

I use M 50's as my overalls. Years ago I went through 26 of them to get 6 good ones when they were readily available. In my library I have three omnidirectional Beyer 48k/24 bit digital mics that are extraordinary. The sonic landscape and imaging is spectacular. It's sort of like watching a Cinerama movie with one camera with a very wide lens instead of three cameras. They don't have the same kind of reach that the M 50's have though.

I run my Decca Tree in the 2 meter by 1 1/12 meter configuration, although I've been subsequently told by various engineers that's not necessarily the correct dimensions.

What do you use for spot mikes, and how do you determine which ones you should use?
The determination is based on the score. If I'm looking for a more aggressive sound because I need to compete against sound effects, I will use different microphones on my strings, like my AKG C 12s. Normally I would use Sony C-37s, which is the same thing I use on the violas, but on the celli I generally always use the C 12s. I try to look at my sweetener microphones as something that is complimentary to my overalls so that when you open them up everything stays within the same color. That way if you equalize in one area it doesn't start to make other things sound weird.

Over the years I've tried to find microphones that were compatible color-wise. As a result, a majority of the microphones that I use are tubes.

Do you EQ much?
Not too much. I use some EQ but it depends on the score and how the stage is responding. I don't use the same microphones in all the rooms either because they react differently as well. As a result, not only do we have the microphone choices on our palettes, we have the rooms as well.

How do you choose where to record then?
I'll talk to the composer to see what he has in mind and that will determine where we go to record. We want the sound of the room to enhance the score.

Some stages are deader than others. Some are a little on the dead side, so I have to use more reverbs to make things sound good. It's almost like rock recording because since you're in a deader environment, you have to create a more live environment. Taking a room that has a lot of room sound and adding another room to it is putting a room within a room and it doesn't always work, so you have to find different kinds of reverbs that work in the room you're in that don't clash.

Do you worry about surround or do you just try to get a good stereo image first?
I don't deal with stereo at all until we make an album. Surround is what I'm always concerned with, but when I'm recording I set up my surround mics so that they all have the same amount of reverb on them and then I turn them off since they can make everything sound bigger than it is and you can fool yourself. I turn them back on for playbacks.

How do you determine where to place those mics?
Again it depends on the score. Sometimes I go the old fashioned way of just sticking two way up in the air really far in back of the room.

Do you have them looking back at the orchestra?
That's what's normally done, but now I don't do that all the time if the room is reverberant. I'll face them toward the rear of the stage to get the reflected sound.

What's your approach to building a mix with an orchestra?
I don't know how I do this but it's developed over the years that I balance all of the microphones out and preset my EQ before I hear a note of music. Generally when the first note of music is played I'm 95% of the way there. After a while you start to know your gain structure and where things should be and how a particular room responds. I make notes about EQ for a starting place, but I can just go.

I would say the same thing about a rock session, where I'd put it all up at the same time and balance it quickly. I know what I want to do on the drums from years of experience. Generally I'll just ask everyone to play at once and listen to the whole thing, then I'll go in and tighten up anything afterwards. I might ask them to play a little by themselves and refine it, but there's something good about getting your sounds all together and defining what's happening as it's going down rather than making everything an individual sound and then putting them together and wondering why it doesn't work.

What mics are you using on the brass?
I only use one microphone, although sometimes I'll put a sweetener on the tuba. Generally I have a special M 49 that I'll put on omni. I use a U 47 occasionally, but I find that 99% of the time the brass pickup comes from the overalls and leakage into the viola and woodwind mikes, which are sitting right there, so that's were leakage comes into play.

It sounds like you use omni patterns a lot.
A great deal. Bones Howe talked in an article years ago about how omni-directional microphones were the best, even on vocals. I tried it and he was right; they just sound more open. Bones was actually one of the main guys that I learned from; not so much from sitting behind him, but from listening to his records and then later getting to know him when he worked at United Recorders as an engineer.

Do you have any advice for someone just getting into orchestral recording?
One thing we didn't talk about, and it's one of the biggest things oddly enough, is balancing a microphone boom so that it doesn't fall over and kill your microphone. I find that amazing because I

see studio setup guys set up a stage where they have microphones unbalanced on a stand, one slight push and over it goes. It's such a simple thing but it's so important.

Did you give away any of your secrets in this interview?
Actually, there are no secrets. I could tell you every microphone that I use and it wouldn't matter because the difference is in I how hear it compared to my contemporaries. We all hear differently. A lot of us have the same microphones and preamps and do things similarly, but it's a combination of how you put it together and how you hear it. That's why in the end there are no secrets.

17

WYN DAVIS

Best known for his work the hard rock bands Dio, Dokken, Foreigner, Bad Company and Great White, Wyn Davis style in that genre is as unmistakable as it is masterful. From his Total Access studios in Redondo Beach, California, Wyn's work typifies old-school engineering coupled with the best of modern techniques.

Do you use the same setup every time you track?
I'll generally choose the same mics all the time and then modify those selections as seems necessary.

Basic tracks these days are pretty much just drums and bass. Rarely do I work in situations where people are going for keeper guitars on the rhythm track date. Occasionally that happens, so I try to isolate the rhythm section as much as possible so there are options at the end of the session to go punch something in without having to worry about leakage.

What's your drum setup?
I'll tell you what my overall approach to drums is. I feel that the drums are sort of like an orchestra in the sense that there are a lot of instruments, so I don't make any attempt to isolate drums from one another or to do anything that would take away from the overall sound. For instance, if you hit the snare, the whole drum kit rings and vibrates. In my opinion, that's a part of the sound of the set that you want to keep, so I don't make any attempt to narrowly focus mics, or baffle things off or anything like that. I just use the mics that I like and don't do any gating or a lot of compression while tracking. I try just to capture the sound of the drums as close as possible to what they are in the room.

I use 87s on the toms and generally a dynamic mic on the snare. Over the years I've taken to using a couple of C12s as overheads. Depending on the kick drum I sometimes will use two mics; a D12 and either an RE-20 or a 421. The D12 has a scooped out response and the RE-20 or 421 will sort of fill that in a little bit.

Do you put them both in the same place?
I usually have the mics about mid-way into the kick. Generally I don't say anything to the drummer about making the bass drum sound good. If the drummer comes in and he has a front head with no

hole in it, I have a cable that I've made that I can slip in through one of the ports. I have sort of shock mount that I'll mount inside and then we'll put the head back on. The most important thing is for the guy to feel comfortable and have the response from the drums that he's used to getting. If you change that, then his performance suffers and you don't get what you're after to begin with.

Do you use something like a 47 FET or something outside the drum?
Well, I have used a 47 FET before, but because the characteristics of every kick drum are different, it really depends on how much fundamental is in it and how empty the shell is. Some people fill their shells up with pillows and some keep their front head on. Some people have a giant hole cut in the head while some people have one just big enough to put your fist through. It really depends on the drum. In my opinion, there are few magic sounding kick drums out there that have everything you want, so you basically have to tailor the mic to the kick drum and figure out which mic is going to represent the best part of the kick drum for what you're after. I'm usually after something that will be at the bottom of the track fundamentally.

Do you use the overheads as cymbal mics or to capture the kit?
I use the overheads to capture the whole kit but with an emphasis to the top end of the set, meaning all the cymbals, hi-hat and accent cymbals. I basically use C12's almost over the toms and not directly facing the cymbals. I put them off axis from each other a bit, so that the two C12's are looking in the opposite directions a little bit. They're sort of close together, maybe a foot or 18 inches apart looking in two different directions back towards the mic stands.

If the intention is for the drum sound to be real ambient, which is the case in a lot of rock situations, I usually put the overheads about 2 feet above the cymbals so they're capturing a fairly wide angle.

Do you mic the hat?
Yeah, usually with some kind of small diaphragm condenser microphone like a 451 or a KM 84. It depends on the sound of the hat and what the guy is going to be playing. If he's going to be bashing on a hat with a real loose pedal, it'll be different that if someone like Vinnie Caliuta is playing and doing a lot of intricate hi-hat work.

Do you use a bottom mic on the snare?
Yes. I rarely use very much of it, but sometimes it really comes in handy. It really doesn't matter much to me what that mic is. It can be just about anything. It usually comes down to whatever's left. If I have a 451 with two 10dB pads available, I'll put it under there.

What are you using on the top of the snare?
I pretty much use a 57 all the time. Occasionally I'll put a 451 on the snare but it has to be the right kind of snare and the right kind of player.

How about the bass? Do you usually just go direct or do you use an amp as well?
Always both direct and with an amp if the bass player has an amp that he wants to use. I put a 67 about a foot away from the cabinet.

How do you handle leakage?
For a modern multitracked recording session where people are planning on going back and having another look at what they've done on the tracking date, it's important not to have a lot of leakage so that anything can be replaced without interfering with something that's on the track.

When I'm at my studio, I've set it up so that there's virtually no leakage. I have sliding glass doors that adjoin the dead room to our live drum room, so the drums are isolated. Then we have a couple of iso booths, so the leakage really isn't a factor.

In situations where the band wants to play and capture the rhythm section as a unit on the spot, I don't worry about leakage. I actually treat it as part of the overall sound and try not to have any glaring phase anomalies.

How do you get your guitar sounds?
It's just a process of guitars, amps and the players. It's trite to say, but so much of it is really in the fingers of the player, so I really work with them and try to find out what it is that they're doing and what it is that they want to capture. On hard rock guitar with screaming Marshalls, the one thing I try to avoid is placing the mic straight on to the speaker. I usually try to be off axis a little bit so that I can avoid the build up of that 1k to 2k screaming, tear-your-face-off sound.

I have the mic back about 2 or 3 inches depending upon how loud it is. Lately I've been favoring this Royer mic (the R-121) for guitar. That mic takes EQ so well after the fact. It automatically shaves off some of that 1k to 2k brittle Marshall thing that really builds up after 4 or 5 tracks of guitar.

Are you using just the one mic on the cabinet?
Yeah, I usually use just one mic close up. I haven't had a lot of luck introducing much ambience into multitracked, layered guitars. It just creates a mess. With more minimalist stuff it's really cool though. I usually end up asking the guitar players to turn whatever tone control they have on their guitars back a hair. It takes just little bit of the edge off. At first they're a little bit hesitant, but there's usually plenty there to go around. It makes it sound a little bigger, especially if you're layering 3 or 4 guitars on top of one another.

When you're layering guitars, are you changing the mic or the mic placement at all?
No, just pretty much changing the guitar. I generally try to use different guitars and different pickups, but I use the same input path for multiple guitar passes.

Do you have any mic preamps that you like in combination with specific microphones?
Yeah I do. Back when Dean Jensen was alive, I bought 12 of those Boulder mic pres that he made. They never really caught on, but I really like them. The only problem is that some consoles can't handle their output on a loud source even when they're turned all the way down, so I've made some passive in-line pads that I can put on those guys. I use Dean's stuff on things with a lot of low end content like bass, toms and kick because of the linear nature of the low end coming out of those things.

I like the old Neve stuff on guitars. The overheads and guitars I'll usually put through a pair of 1073's.

Are you compressing while you're recording?
I usually don't add any compression on the tracking end of things. I try to maintain all the dynamic range that's there because I find that later it leaves me a lot more options about how I want things to sound.

What's the hardest thing for you to record?
Somebody that can't play very well (laughs). Truthfully, that's a lot of people nowadays. The art of being prepared for the studio, along with a lot of the engineering arts, is being lost in all the cut and pasting. I've found that the preparation that people have before coming into the studio has diminished over the last few years by an astounding amount. People will come in and work hard to get something on the first chorus and then say, "OK, can't you just paste that everywhere now?"

When people used to play these performances from top to bottom there was a synergy with the track that happened. Something would evolve as the track went on. You definitely lose that if you're just using a hard disc recorder as a glorified musical word processor.

How do you approach vocals?
Vocals, on the other hand, I do compress going to tape all the time. It depends on the vocalist, but I'll use any number of mics. It's almost always some kind of condenser mic and some kind of tube mic. There's a lot of really great vocal mics out there that do a great job; it just depends on who's singing. It can be any number of microphone preamps, depending upon who's singing and what kind of sound you're looking for. For tracking, I use an LA-3A with a quick attack, slow release, letting that lightly catch anything jumping through.

Are you looking for something that sounds good by itself or something that fits in the track?
Usually something that works in the track. If it's a ballad where the vocal is going to be way out in front and has to stand on its own, I'll just be looking for a good vocal sound, but usually I'll make adjustments to make it work with what's being played back.

Do you send the same FX to the headphones as what you think will be used on the final mix?
I do it as requested, but I generally try to keep the headphone mixes pretty dry. I want them to be punchy and fat and basically in their face because I think it keeps everyone really honest. If somebody wants some verb or delays, it's not a problem, but I try to keep it down to the very least that they'll accept. I'll slide it in there and keep on asking if it's enough and explain to them that I prefer that they just go ahead and sing it sort of au natural.

Do you use room mics on drums?
Depending on where I am, I really love M 50's. I don't own them but I'll rent them. In my drum room I'll velcro a couple of PZMs to the wall that the drums face and use those as room mics sometimes. I'll

also use a couple of 87's sometimes. If I use 451's on the overheads or an old set of 414's, then I'll use my C12's as room mics.

Are you EQing when you record?
I do whatever it takes to make it sound the way I want it to sound. Generally I'll start with the microphones, but then I'll do whatever I have to. With a really good studio drummer, there's very little that has to be done because the kit will sound great right off, but that doesn't happen that often. For the last half dozen years or so I get the guys from Drum Paradise to bring some drums or tune the ones that are there if there's a budget that can accommodate it.

Do you have a philosophy about how you record?
The overall philosophy is to make everybody as comfortable as possible. In a tracking situation, aside from your responsibility of getting something decently recorded, the most important thing is to get good headphone mixes for these guys. In fact, to get the best one possible. Amazingly bad things happen to even the best players when the headphone mix is all screwed up. I don't relegate the headphone mix to anyone else. I make sure that I have a set of headphones that I can switch across all the cues that are being fed to the guys playing in the studio.

Beyond that, after almost every take I will have an assistant make a sweep of every headphone position and listen to a playback to make sure that none of them have gone south or an amp is starting to distort or something like that. I don't think you can pay enough attention to that part of it because if the guys are hearing something that feels good, it moves the session from sort of a technical exercise for the musicians to a real inspiring and fun thing. When you can create that atmosphere in the studio for them, that's when you're going to get the best work out of the players, and when you get the best work out of them, it's going to sound better. It's really amazing how no matter what tools you're using, if people aren't having a good time, it's just not going to work.

18

EDDIE KRAMER

Unquestionably, one of the most renowned and well-respected producer/engineers in all of rock history, Eddie Kramer's credits list is indeed staggering. From rock icons such as Jimi Hendrix, The Beatles, The Rolling Stones, Led Zeppelin, Kiss, Traffic and The Kinks, to pop stars Sammy Davis Jr. and Petula Clark, as well as the seminal rock movie "Woodstock," Eddie is clearly responsible for recording some of the most enjoyable and influential music ever made.

How did what you learned at Pye Studios in London when you were starting out shape your philosophy about recording?
In regards to mic techniques, what I adapted was this classical idea of recording; i.e. the distance of the microphones to the instruments should not be too close if you wanted to get anything with tremendous depth. Obviously I used close miking techniques as well, but it started with the concept that "distance makes depth" that my mentor Bob Auger taught me. Generally the basic philosophy was getting the mics up in the air and capture some room sound and air around the instrument, then you'd fill in with the close mics.

Of the microphones that we used, 67's were probably the favorite (and still are today), but we used 47's, 251's, a lot of KM 56's and 54's, ribbon mics, AKG D-12's, D20's and D30's. In fact, on some of the Hendrix stuff I used a D30 on the bass drum, which I still think is a really great bass drum mic.

Once I came to the United States in '68, utilizing that philosophy seemed to work, but with some modifications. Obviously watching how the American engineers did things influenced me to a certain extent.

How was that different?
It was different in that they didn't use as many mics and they would be very tight in, which I though was a cool thing, so I adapted that close-in technique of getting right in on the speaker cab which seemed to work very well.

Were you using a combination of close and far mics?
Yes I was. In fact on the Hendrix stuff in '68 at the Record Plant on the *Electric Ladyland* album, if you listen to "Voodoo Child," you can hear the way the room just resonates. That's because I had mics

everywhere, and the fact that Jimi was singing live too! I wasn't afraid of recording an artist in the room live as he was cutting. To me, anything that was in the room was fair game to be recorded. Don't forget that I had an artist who was an absolute genius, so it made life a lot simpler. When you're recording someone of Hendrix's ilk, you're not going to be overdubbing much if it's a live track. You put the mics up, place them correctly, and give the artist the room and the facility to work in and make sure it sounds cool so when they walk into the control room they say, "Oh, that sounds just like I was playing it out there." That's the goal; to capture the essence of what the artist is actually doing in the studio.

Obviously there are other ways to do it. You can do it in sections and pieces by overdubbing and recutting and that certainly works too, but to me there's nothing more exciting that having the band in the studio cutting live straight to tape where that's the performance and that's what gets mixed. That's the essence of any great recording. I don't care if it's classical or rock or country, you've got to capture that performance and the hell with the bloody leakage.

Too bad that DAW's have changed that these days.
It has and I think to music's detriment. I strongly feel that music should be captured as it's going down. If you make a mistake, too bad. You cut another piece and chop it together, but you still have the essence of that live performance.

So you mostly did multiple takes and then chopped together a good one?
Yeah, absolutely. Chopping multitrack tape was the name of the day. I think that a lot of producers and engineers that grew up in the '60s and '70s still hold to that philosophy. I think that even today with Pro Tools one can still do that although it also can be slower in the long run.

When you started you were pretty limited by the number of tracks and channels available.
Definitely. You have to use your imagination and think really hard about how to plan it out. For instance, on Hendrix's stuff, which is the classic example, it was done on four track. On *Are You Experienced* we used mono drums and mono guitars and so forth. We would fill a four track up then dump it down to another four track, leaving two tracks open, then you may have to do that again. On *Axis: Bold As Love*, I was recording stereo drums which made a big difference.

Was your approach different when you went to stereo?
Yes. When it was mono I just used a single overhead, a snare mic and bass drum mic. There might be one or two tom mics but that would be it. When I went to stereo I probably used a pair of 251's or 67's, I can't remember which. I was just trying to get that left to right image when the toms would go left to right. I always record from the drummers perspective and not from the listeners perspective.

Has your approach to tracking changed when you do it today?
Yes, it has been modified in the sense that you don't have to use an enormous room to record the drums anymore. In fact, bands today don't want that huge reverberant drum sound that we used to love, so you can record drums in a smaller deader space and still get a big fat sound. Obviously I'm

using more mics, multiple mics on the bass drum, multiple mics (top and bottom) on the snare, which I didn't do before. I use a lot of mics on the guitar and then pick the ones that I like.

Is your setup the same all the time?
Pretty much. I will experiment with different microphones as they come in. The [Shure] KSM-27 is a great guitar amp mic. I love the new KSM141, which is a cross between a 451 and a KM84, on hat, percussion, acoustic guitar and underneath the snare. The SM91 and SM 52 are my bass drum mics of choice, and I use KSM44 on overheads, but I still use vintage mics like 47's, and the new Neumann TLM 103's, 147's and 149's.

To me a microphone is like a color that a painter selects from his palette. You pick the colors that you want to use, so the mics are my palette. In the end it doesn't matter to me too much. Whatever is available, I'll just look at it and think, "I wonder what this will sound like on the guitar, or bass or whatever instrument." I know what my standard stuff is and if I need to do something really fast I'll always go back it, but I'll often experiment with whatever happens to be in the studio.

Do you tailor the mic preamp to the microphone? Do you have certain combinations that you like?
No, just blanket it with vintage Neve modules, either 1073's or 1081's. I like the 1081's because of the four band EQ so I can carve things out particularly when I'm recording bass drum. Lately I've been using the new Vintech X81, which is a copy of the 1081.

So you're EQing during recording?
I always do. I have done so my whole life. If I hear a sound that I like then it immediately goes to tape. If it's a guitar, then I'll print the reverb as well on a separate track so the sound is there and locked in. I usually have an idea of what it's going to sound like in the final analysis so the EQ and compression is done right then and there. I think if you bugger around with it afterwards you have too many choices. This isn't rocket science, it's music. Just record the thing the way you hear it! After all, it is the song that we're trying to get and the guy's emotion. We're becoming so anal and self-analytical and protracted with our views on recording, I think it's destructive and anti-creative. It's bad enough that we have to be locked into a bloody room with sweaty musician (laughs).

Recording music should be a fun filled day. To me, making a record should be about having a ball because it makes the day go quickly and yet you're still getting what you want on tape. There's a friend of mine that has a bar in his studio and after the session is finished everybody has a beer and relaxes. What a wonderful thing! I think artists today have a tendency not to do this. You cut to a bloody click track, go to Beat Detective, do a lot of overdubs in Pro Tools, and then spend a lot of time searching for the right plugins to make it sound cool.

But the track has to move and breathe. Listen to all the great songs and albums that have been recorded the last 30 years. The ones that really stand out are the ones that breathe and move. With human beings, their tempo varies. I do admire what can be done in Pro Tools, but if there's something that wrong, you should have done another take and maybe chop things together.

What's the hardest thing for you to record?

The toughest thing to record is a full orchestra. Getting the right room and properly placing the microphones is really tough, but it's also so rewarding. The other thing that's tough is the artist that can't get the right feel so you have to go through a lot, changing microphones and instruments and placement, to make it work. That can be boring.

I like to think that going into the studio is a challenge. What usually happens is that the artist, unbeknownst to himself, has done a brilliant job on the first take and it all goes down-hill very rapidly after that. The reverse can also be true in that the first take is weak because the person is just getting used to it and they build up gradually to point where it "is" great.

19

MACK

With a Who's Who list of credits that includes Queen, Led Zeppelin, Deep Purple, The Rolling Stones, Black Sabbath, Electric Light Orchestra, Rory Gallagher, Sparks, Giorgio Moroder, Donna Summer, Billy Squire, and Extreme, the producer/engineer who goes simply by the name Mack has made his living making superstars sound great. Having recorded so many big hits that have become the fabric of our listening history, Mack's engineering approach is steeped in European classical technique coupled with just the right amount of rock & roll attitude.

Do you have a philosophy about recording?

I try to get the biggest, pristine sound that I can so it can be bent in any direction later. Something small and tiny is really hard to make bigger. You can always screw the sound up later after you've recorded it (laughs). Sure, you can say "OK, this requires a small sounding piano" or something like that, but you're confining yourself and you can't change your mind later.

That goes for multitrack recording, which in the old days, if you had to put a band down on an eight track machine then you'd record them on two tracks and have six left for overdubs, so you had to have a precise image of what the balance needed to be when you started recording.

Do you have a standard setup that you start from every time?

No. It's totally dependent upon the type of music. Different types require different setups. If it's something with a really fast tempo, you would mic things tighter than if it was a slow bluesy thing, which is better with some open space. I would pick the microphones and placement of the mics with that in mind.

How long does it take you to get things where you like it?

Probably anywhere from 20 minutes to an hour or so. I tend to work really fast. I don't want anything technical to get in the way of the music. You usually don't get a lot of time anyway because people are frequently wandering around and anxious to play. I like to use that time to get the whole setup done when the players are pretty uninhibited. When we start taking [recording] I don't want to interfere with the creative process and go, "Can you give me that left tom again, and again, and again?"

That doesn't give you much time to experiment.
Not all that much but I get that time back because it's inevitable that the band will go through a song and come to a passage where they want to change something. While they run over things again and again, that's when I use the time to check individual things out and experiment.

Do you use your overheads as the basic sound of the drums or just as cymbal mics?
The basic drum setup would be bass drum and overheads. My favorite would be B&K's but I like to use Schoeps if it's not a hard hitting drummer. Everything else is there to augment that sound.

Do you put the overheads over the drummer's head in an X/Y configuration?
No, I try to make sure that they're an equal distance from the snare. It does depend on the room. In a huge room I might use an X/Y thing but the rooms for rock stuff are usually on the smaller side so I use an A-B.

Are they pointing straight down or at the snare drum?
They're pointing directly at the snare drum.

What do you do with the kick?
I use two mics; a close one and one far away. I use something like a D12 up close but a little off axis angled downward, depending upon if you have a front skin or no front skin. I use a U47 for the far mic about three feet away but very close to the floor.

I really like to use my own microphones because I know what they sound like. Even though a mic might have the same label, it still might sound different, so I like to use my own because I know what they do.

If your main drum sound is coming from your overheads, what are you looking for from your other mics?
Actually apart from the close kick mic, which is a dynamic, everything else that I use are condensers. For example, I use 67's or 87's for the toms and something like a KM84 or a AKG 224 for the hat. Probably what's really different, because I haven't seen anyone doing it except really old guys, is I put the snare drum mic exactly parallel to the drum.

Pointing at the side of the drum or pointing across the top?
No, pointing directly at the drum. That's a very old fashioned, classical drum recording technique.

Are you pointing it at the hole on the drum?
No, because that tends to cause the occasional wind noise.

How far away?
About 10 to 12 inches away. I like an AKG 414 in hypercardiod. Ideally I would like to use every mic in omni because they sound best that way but you can't always do that.

You've done so many guitar bands with great guitar players, what is your approach

to recording electric guitar?
Just leave enough distance from the amp so you get a bit of room reflection to it. I used to do the thing where you crank the amp so it's noisy, then put on headphones and move the mic around until you find the sweet spot. I usually use two mics (which is sort of contrary to my beliefs because you get a lot of phase stuff) because you get a natural EQ if you move the second one around. If you can remember what the hiss sounded like when you had a good guitar sound then half the battle is won.

One of my big things is not to use EQ, or as little as possible, and not to add any but find what's offensive and get rid of that as opposed to cranking other stuff to compensate.

What mics do you use on guitar then?
I like a KM84 and an SM58. One is straight on axis and one is off to the side.

Does it matter where the amp is in the room?
Yes, that matters very much. It's the same philosophy as with a monitor speaker. If you pull it away from the wall by a foot or two then your whole system sounds different and the same thing applies to guitar amps. Little things like tilting it a bit or changing it around. For some reason amps are usually put in place by somebody like a roadie and nobody ever moves it after that, but moving it around a little and angling it can really make the sound change a lot.

Do you usually have everyone playing together trying to get keeper tracks?
I try to get everybody at the same time. I recently worked with Elton's band and everyone was like "Wow, he's letting us all play together in the same room. This is pretty cool."

You don't care about leakage then?
I do, but there are gobos and blankets to help out. If it's a good band then you do notice the difference. Stuff that has been layered in parts are just not the same. The little accelerations and decelerations are so together that it just makes things come to life. I'd rather leave the little flaws in or repair them latter. You don't notice a lot of them any way. It's the performance that counts.

I try to keep everyone pretty close so they can communicate outside the headphones. There's nothing worse than putting someone in a box out of his environment.

What do you do for bass?
My first thing is direct. I do record the amp just to have it, but unless it's really good, I don't use it. I prefer small amps to big ones. The big stuff never really does it. For guitar amps, Marshalls are pretty standard but with everything else, smaller is better.

What do you use for mic preamps? Do you have a certain combination that you like for certain instruments?
I totally sold on Millennia's because I think that transient response needs to be the best that it can and I like the cleanest possible sound to get it. I don't want any extra ingredients. I just want the recorded

sound to sound the same way it sounds in the room.

Do you use the Milennia's on everything?
I use as many as I can get my hands on. Neve's are good too. They have a certain sound that I can deal with. The actual sound of something is mostly determined by the initial instant of the sound. If you cut that off, then it could be any instrument. You can't tell what it is any more, so I got really hooked on preserving the transients.

What do you usually use for piano?
My favorite mics are the Sennheiser shotgun mics, the MKH425, in X/Y. It's totally inappropriate and I know that, but it really, really works. I never have to do anything other than put them in the piano.

Where do you place them, where the strings cross over?
A little lower than that. They're about 5 inches off the strings. It depends on how hard the piano player hits the keys and what range he's playing in. I was forced to do this one time because there was a really hot amp right next to the piano and I had to get in really close. All of a sudden I realized "Hey, this is better than anything I've ever done," so I stuck with it.

How do you record vocals?
I like 47's. Just for the heck of it, I once had 10 new 414's set up against one another with a willing singer, which is usually a problem because if you have too much of a Christmas tree set up people get intimidated. It was unbelievable. It sounded like you were putting in various filters from one mic to the next. They were all supposed to be the same. I found that experience shocking so from that point on, I always carried one mic for vocals that was not used for anything else.

Are you using a Millennia for vocals as well?
I always use the HD3C Millennia with the built-in Apogee converter. I've had it for about eight years. I come straight out digitally to whatever I'm recording onto. I use a Manley Vari-Mu for a compressor.

Do you compress much while recording?
I do compress the bass with like an 1176 or the Manley by about 6dB or so to keep it tight. The better the bass player, the less you need it. You want something that has a slow release time so it's not pumping.

What's the hardest thing for you to record?
I don't like doing vocals with people that can't really sing. That's probably the most tedious thing for me. Also, I'm not that good of a liar. I have a hard time not being honest, especially when you know from the first take that a vocal is going nowhere. But with people who can really sing doing vocals is not that big a deal. With Freddie Mercury, you'd know that you'd be done within the hour. He'd do a few tracks that would be great and then just leave you to put it together.

20
SYLVIA MASSY

Grammy-winner Sylvia Massy has worked with a wide variety of artists across many different musical genres. These include Tool, Johnny Cash, Red Hot Chili Peppers, System of a Down, Prince, Sheryl Crow, Tom Petty, Smashing Pumpkins, and Lenny Kravitz, to name just a few.

Always willing to think outside the box when it comes to recording, Sylvia's book *Recording Unhinged: Creative and Unconventional Recording Techniques* outlines her experiments with processing sound through light bulbs, household appliances, and even food.

She and manager Chris Johnson are also owners of the largest microphone museum in the world, with over 2,200 vintage pieces, including many rare prototypes. You can see example on her website at sylviamassy.com.

Is your recording approach different for different styles of music?
Absolutely. I'll choose the microphones and the studio arrangement according to the type of music. Almost every session I do will be slightly different. For instance, a project I did this past weekend was an Americana project and we worked with no headphones and the two singers sang together on one figure 8 mic with them standing on either side. Typically I would never do that but everyone was playing in the same room.

If I was doing something louder like rock then I'd want to isolate the amp cabinets so that the spill doesn't get into vocal mics or drums. It helps when it comes to adjusting the sounds later in the mix.

How long does it typically take you from the time you get the mics set up to the time you begin to record?
On a brand new session it will typically take from one to three hours from load in to get everything ready. By hour four we're fully up and going. It's easier when you have a studio where you have everything that's working [laughs] so you don't have to troubleshoot a lot.

So I have that four hour mark where will be in the midst of recording. There's that moment in the session when you're in the boat at the dock and you're pushing your way into the deep water. That's when you're concentrating on the music and the performances rather the technical, and that's when I consider the session as going.

Do you have a typical drum setup that you use?
I'll have a mic in front of the kick drum, one on the top and bottom of the snare, then mics on the top and bottom of each of the toms, a mic on the high hat, one on the ride cymbal, then a pair of overheads or a stereo mic above the kit. The models of mics that I use will change from session to session.

Generally I'll gravitate to use the microphones that I know will work in all situations like SM57s, 451's U87s, so I'll go there first. If there's time I will experiment.

One of the most exciting things I'll do (if I get a chance) is to use speakers as microphones on the drums. I'll reverse-wire an old woofer and put it in from of the kick drum for a deep subsound that I'll blend in with the normal kick mic.

Are you using room mics?
I love to use room mics because I can really get crazy and fancy with that. Generally I'll use some AKG 414's because they're easy to use, and then I'll add several "unusual" room mics that I'll treat differently. Each of these mics will have their own character but then I'll treat them with compression and EQ to make them even more unique.

One of the things I love to do if given the chance is to take old boom boxes and cassette recorders that have a microphone in them, and then I'll wire them up so that acts as a microphone also and record it to the DAW. I don't want to depend upon the experiments because some of them don't work out, but sometimes it's really exciting, especially with old cassette machines, because they have built-in compressors that are wild and really adds to the sound of the drums.

What do you usually try first for overheads?
Most recently I've been using a pair of the Mojave MA-200s, but I just purchased a vintage AKG C24 which is a stereo version of the really famous AKG C12. This is actually the same mic that I used on all those great Sound City recordings. The mic has two capsules and you can adjust them so you can change the directionality.

For instance, when I worked on the Tool record I used the C24 in an MS configuration, but over the weekend I used it in an XY.

So you're using the overheads for the sound of the drum, then filling in with the other mics then, right?
Yes. A lot of engineers will concentrate on having the center of their overheads the snare drum, but I'm capture the drum kit and especially the cymbals. The cymbals may not be in perfect alignment around the snare drum, but I'm concentrating on capturing the best picture of the drum kit in its entirely.

You mentioned that you use underneath mics on the toms as well.

The way to use the two mic technique on the toms and snare is to have the bottom mic flipped out of phase. Then you have to listen carefully to how they are aligned with each other. The top mic is going to give me the attack, but the body and the tone of the drum is better captured with the mic that is underneath. I'll bring up those two mics on the desk [console] and then I can adjust the level of the bottom mic to fill in as much low end and body in that drum.

I'll typically take the top and bottom microphones and sum then to one channel in my recorder. The same with the snare. I like to commit! We used to do that on tape because we only had a limited number of tracks but in fact I find that keeping those things separate now causes more phasing problems later. It never sounds as punch and as strong as if you combine them before you record.

You mentioned EQ and compression before and I take that to mean that you're not afraid to EQ or compress while you're recording.
Yes. I want to record the song so that it sounds finished when it goes to the recorder for a couple of reasons. It's easier to mix because you don't have to try to recreate how great it sounded while you were tracking. It's super easy when you're mixing then. You just push up the fader and it sounds great.

The other reason to do it that way is that you may not be mixing it, and this is a way to guarantee that whoever is going to mix isn't going to screw up your work. They're going to get it right because you've already committed.

Are you from the school that if you put the faders all up to zero then your mix comes out?
No, because I find that whatever's being recorded into the DAW has uneven levels. Certain tracks are recorded at very low levels. When I get sessions like that I have to gain up those very low tracks in order to get the juice I need for the mix. So no, those low levels don't work for me.

How do you handle leakage? Do you embrace it or try to get rid of it?
It depends on the project for how the bleed between instruments will work. Sometimes I love having a live vocal in the tracking room when we're laying down basics because the energy of the singer in the room will make everything else sound better or real. Sometimes you'll get a great performance from the singer while they're standing in front of all the other musicians. I try to use a mic that will be as directional as possible to reject a lot of the other sounds of the room.

A lot of it depends on the players as well. If you have a guitar player that's clamming it up and those mistakes are being recorded into the drum mics, then you're going to have a hell of a time later when you mix, so I will usually start by isolating things.

If you're doing Americana like my last project, then you want to get the bleed and everything mixing together. If there are great players then you don't want to miss any of that, so let it bleed.

How about mic preamps?
Honestly I think the choice of microphone preamplifiers is maybe the most decision to make in any

kind of recording chain, because you can have inexpensive microphones but if you put them into cheap mic pres then it will very difficult to get a good recording.

When possible I'll always try to get my favorite mic pres which are Neve 1073s. That module seems to be the magic bullet for any instrument. It makes things easy because there are so few frequency choices for EQ and they're all good, so you can't really screw it up.

How do you approach recording acoustic guitar?

For acoustic guitars I like to use one mic. There are two types of acoustic guitars that you can record. There's the more rhythmic percussive sound that lays into a track for texture, and then there's more of the melodic type. I feel that for the rhythmic acoustic I'll use a dynamic mic just to get a nice woody tone. For the melodic type I'll use a high end condenser. One of my favorites would be a U47 right up on the body of the guitar with just a little bit of compression.

Let's talk about electric guitar. There are a lot of different approaches. What would be yours?

I like to have two microphones on the cabinet. I'll choose one speaker and put both mics close to the grill cloth at the same distance. It will usually be a Shure SM57 and a Sennheiser 421. The 57 will give me the edge and the 421 will give me the body. I'll blend to taste and then I'll combine those two mics together to one track.

I usually don't use room mics unless it's for a solo part, then I'll have a condenser or a ribbon mic out a little bit in the room. I'll record that on a separate track. Usually I like to commit to a single performance on a single track.

How about acoustic piano? Is there a general starting point for you?

My favorite recordings are with a pair of M49 tube mics. I usually will put both mics up close to the hammers, and one will be on the high strings and the other will be on the low strings. They may be in an X/Y position so they're not cancelling each other out.

One other technique that's cool is to put a PZM microphone taped underneath the lid, and that gives you a great overall picture of the piano.

I've also used another technique that I've also used on drums as well. I put a mic into a garden hose and then wrap it around the harp of the piano. That captures a real deep intimate sound of the piano that I will blend in with the other microphones. It's unusual and not for everybody, but for certain artists it gives sort of an unusual creepy tone which is great.

Now if you're doing upright piano that's a whole other story because it's much harder to get M49s down into the piano. I'll pull the piano away from the wall and then mic the soundboard at the back. I'll use a couple of condensers, with the same idea of miking the low and the high strings. If it's a banging sort of Beatles thing I might just put one mic on there and call it a day. It could be a dynamic and it will be great.

Let's go to vocals. Do you have a particular mic that you know is always going to work?
I find that U47 will be great for both male and female voices, but I'd choose a C12 for a male voice because it's a bit brighter and on certain female voices it's too shrill. I go for a deeper, warmer mic for a female voice. The Neumann M49 is great for both a female and a male voice.

However, you can simplify all this by just a plain old SM57 or 58, which is also a brilliant mic for vocals. It's been proven to work over and over. We did a shootout with a U47, an M49 and even an ELAM 251 (that's about $100k worth of mics) up against an SM58. We had Billy Corgan from the Smashing Pumpkins come in and sing the same verse of a song on each of these mics and then we blindfold tested as we listened to each track without knowing which mic was used for the recording. It came down to either the ELAM 251, which is now a $35,000 microphone (if you can find one), and the SM58! The advantage of the 58 over the other mics is that Billy could hold it so he could get a better performance. I could say that's true for any singer, especially for a rock vocalist who wants to grab it.

Imagine a punk rocker trying to sing into a mic suspended in front of him as opposed to a mic he can hold. You're going to get a much better performance, and if the sound is not that far off from the $35,000 mic, then go ahead and use it.

I know you're a big collector of microphone who's traveled the world looking for vintage units. You've studied their history extensively. After you've done it all, was there one thing that jumped out that you didn't know before you started on your journey?
I can tell you one very profound thing is that microphones are an important part of the history of technology in the modern world. The microphone in the first telephone 1876 started this whole electronic revolution that we're still in. If it wasn't for the microphone, we wouldn't be talking to one another right now over Zoom. It all started with Alexander Graham Bell's microphone.

The collection of 3,000 microphones that is here represents so much history. The people that had the microphone had the power to speak to large audiences throughout modern history had the power. You'll see it again over and over. The loudest voice has the power and it's all because of the microphone.

21

DENNIS MOODY

Known as "the drummer's engineer" for his smooth, natural drum sound, Dennis Moody has been the choice of top-shelf session drummers like Dave Weckl, Steve Gadd, Michael White, and the late great Ricky Lawson when it comes to their own personal recording projects. Dennis has also worked on projects with major artists like Prince, Aerosmith, Diana Ross, Iggy Pop, Smokey Robinson, Bill Medley, Missy Elliott, Mike Stern, Randy Brecker, and Miles Davis, just to name a few.

Dennis is also one of the few studio engineers who regularly crosses over to live concert engineering, having mixed Front Of House in venues like Carnegie Hall, Madison Square Garden, The L.A. Forum, Wembley Arena, and Royal Albert Hall. He has also mixed thousands of shows of Broadway style musicals, large and small orchestral shows, plays, and live broadcasts and podcasts worldwide.

Do you have a philosophy about recording?

Yes, I do. Keep it simple and keep it real. I think that artists are polishing things so much these days that it takes the feel out of the music. Who cares if you use a vintage U47 or an SM57 as long as it works. Of course I'd like the 47, but we use what we have and as engineers, we have to make things work.

I recently had a very well known singer in the studio and we used an SM58 on him for a scratch vocal track. He went home and tried to beat it with his vintage U47 and couldn't match the performance. We ended up using the live tracking vocal for the record and it sounds awesome.

That's why I like to do all the jazz projects that I do. They've got a different vibe. To me, keeping it simple, natural, and not overdoing things is the way to go.

You do a lot of different styles of music. Is your approach different for each?

No, I pretty much do things the same way. I use the same drum mics, although for metal I might change the kick mic. For most styles I'll use a subkick. For straight ahead jazz I won't.

How long does it take you to set up a session and be ready to go?

Once the mics are set up, I can get rolling in about 45 minutes. Maybe an hour at most. Drum sounds I get in 15 minutes now because I know what I'm looking for. I don't need to experiment or see what the options are. We may mess around with the tuning, and maybe tape up some heads a bit, but I try to keep it basic.

Let's talk about drum mics then. I know that your setup has changed over the years that we've talked about this, but what is it now?
Beta 52 on the kick because it has just the sound I'm looking for. SM57 for the snare always, with no bottom mic unless the snare is more than about 10 inches deep. Even then I rarely use it.

For high hat my favorite is a vintage KM84, not the new KMi84 because they don't sound like the vintage ones. I don't have one in my studio so I've found this Studio Projects C4 that cost me about $100 that has the sound I'm looking for. It's not too edgy on the high end, and has a nice full sound. I even used it on the [legendary drummer] Steve Gadd-album I recorded a few years ago with great results.

For overheads I was using a pair of 451s but I've since picked up a pair of [AKG] 460's. They have more body to them and sound great. For toms, if I had my choice I'd use the Josephson E33s. They're great mics but buying 4 or 5 of them is a bit out of my budget. Here in my studio I use the [Shure] KSM141s for the rack toms, and on the floor toms I use the KSM 32s. They have bigger diaphragms and deliver a bit more body on the larger floor toms.

For a room mic I'm using the Shure VP88 stereo mic. It took a while to get used to, but I really love it. I found that in my room, the further I move it back, the stereo imaging really becomes more apparent.

Are you trying to capture the entire kit or just the cymbals with the overhead microphones?
Mostly the cymbals and I have a reason for that. I used to place the overheads higher until I did the Gadd record. Steve likes to hear the kit like it sounds to him when he plays it. He hears the cymbals the loudest, so that's where I placed them in the mix. I played a rough mix of the album for [former Journey drummer] Steve Smith one time while we were riding somewhere on our tour bus. He said, "That's awesome. That's the way I hear the cymbals when I play."

If you have the overhead mics placed high then you'll probably need a ride cymbal mic. If you need to raise the ride when it's played when only using a pair of overheads, you can hear the balance of the kit change due to the leakage from the other drums. So I put them 12 to 14 inches from the bell. The reason why they're over the bell is so you don't get the swishy sound of the cymbal as it rocks back and forth. Blending in the stereo room mics then pulls the whole drum sound together.

The side with the ride cymbal usually has a crash so I'll place the overhead in between them. If I need more ride, I'll just move the mic a little towards it. Of course I can add an additional ride cymbal mic, but I rarely ever need it.

Where are you placing the snare drum mic, considering that you're not using an under-snare mic?
It's almost parallel to the top of the head with a little tilt towards the center of the drum. I put it two

fingers above the rim and just where the capsule meets the body of the mic. I bring it in as close to under the high hat as I can so there's more hat rejection. Understand that I have to EQ it quite a bit to get that snare sizzle to come out, but it always works. I'm hearing far too much bottom snare mic in popular mixes these days. To me, it's a very unnatural sound.

What are you using for preamps?
For the kick and snare I use a GML 8200 both into an API 550A EQ. For toms I use the Sunset Sound S1P "tutti" mic pre. This is a modified API design that really opens up the sound. Having grown up with API consoles, I really love these SS mic pre's.

Does that mean you're EQing while recording?
Absolutely. I always EQ "to tape" as we used to say. It's the old school way of recording, when you'd have to record drums to four tracks. You had to get it right. I try to get it sounding as close to final going into the DAW. Some projects I track, I don't get to mix, so I want whoever is getting the project next to put those faders up and go, "Wow!" Additionally, I use very little compression while recording as it can not be effectively removed once it's recorded that way.

How are you miking the acoustic bass?
I try to record acoustic bass on two channels using a pair of Avalon M1s [mic preamps]. I've been using a [Sennheiser] 441 lately placed about one foot away and at the bridge. I want to get the finger sounds and string noise that naturally comes from an acoustic bass. On the other channel, I take a DI from the acoustic bass pickup. These days there aren't many bass players that don't have a pickup or a small clip-on bridge mic installed on their bass. I'll use the mic channel and then add a little of the pickup until it starts to sound artificial, then back it off a bit. This gives the bass some midrange and adds to it's core sound..

What do you do about leakage?
I try my best to get rid of leakage, but I embrace what's there. When I record a piano with a vocalist or an acoustic bass in the same room, I use a pair of DPA 4099's over the inside of the piano and close the lid and put a blanket over it. There will still be a bit of leakage from the piano into the vocal mic, but there's very little vocal on the piano. If there's leakage I just deal with it and use it to enhance the live performance sound.

How do you approach acoustic guitar?
I like to use just a single mic in an ambient room. AKG 451's and AKG 460's sound great, as do the KSM 147s. I back it off a foot or 18 inches and tell the guitar player not to lean into it. If it gets too boomy I move the mic back to reduce the proximity effect and the low end build up.

How about electric guitar?
An SM57 on a Marshall usually delivers a great sound. A [Neumann] U47FET also sounds good. If it's a

rock thing I try to move the mic back another foot. I place it about half way between the speaker edge and voice coil "tweeter cone" so it's not too buzzy or boomy. I don't use room mics any more and it doesn't seem to make a difference. It's easy to add digital ambience later to dial it in to exactly fit what the track calls for.

What are you using for vocals?
I've been using 414's lately because they work really well in my room. I use a dbx 160a with a 2 to 1 ratio and just tap it a little. Sometimes with a female vocalist I'll add a little 250Hz to give her more warmth. I record these using my API SSl "Tuttii" mic pre's.

What's your setup for horns?
My favorite horns mics are U87s. I'd use them on everything if I could. I might use a FET47 on a trombone or even a regular U47.

What's the most difficult thing for you to record?
Low level stuff like a gamelan. My studio is in downtown LA so sometimes I'm picking up stuff that I never even knew was around [laughs]. Harp is another one. They're difficult because the sound is coming from everywhere. I use two mics with my GML mic pres, which you can crank up really loud because they're so quiet. I pull those out right away when there's something super low level like that.

If you had only one mic to use, what would it be?
I'd have to say a U87 because it can sound good on just about everything. Back in my days working with Motown when I was first starting out as an assistant, I used to see some of the engineers use an 87 on a kick drum. These days many of us use a 47FET on a kick so that shouldn't be that much of a surprise.

Have you found sonic difference between the various versions of 87s?
Yes. The newer ones have a different midrange that's a little less pleasant to my ear after using the vintage 87's. Some of the clones sound pretty good. The Warm Audio WA87 sounds really nice, with just a little bump in the midrange. That little added bump actually works great on piano too.

Any words of advice for engineers just starting out?
Just because your favorite engineer is using a certain type of microphone doesn't mean you need to use the same one to get a great sound. You should try everything you have at your disposal and you might be surprised what you come up with. Don't say, "I can't be successful because I don't have the right stuff." The right stuff is your ears and your creative abilities. I encourage new audio engineers to try something different and to be unique. This is the only way you will really stand out from the crowd and develop your own personal style, and not just sound like everybody else.

22

BARRY RUDOLPH

Barry Rudolph has worked with music legends like Rod Stewart, Lynyrd Skynyrd, Hall & Oats, and Pat Benatar, among many others, and is credited with 30 gold and platinum albums. What makes him unique is that he's written over 6,000 audio gear reviews for Mix Magazine, Music Connection and Resolution and others over the years, so he's played with more gear than most people ever will in several lifetimes. You can read most of his reviews and much more on his website at barryrudolph.com.

Is your approach different for different styles of music?
There might be some different mics or locations, but generally not too much. I use the usual things - two top snare mics, a bottom snare mic, maybe a couple of kick drum mics - all the usual tricks that you pick up along the way.

How long does it take you to get a session going?
If the mics are all placed and working, I can usually have the sounds within the first run through. Don't forget that you're also getting headphone balances and things like that at the same time.

What drum mics are you using?
I like to use a condenser and a dynamic mic on the top of the snare. My ultimate condenser choice for this would be something like a Sony C-37A with maybe a Shure Beta 58 alongside. I also like to use a Granelli Audio 90-degree elbow on a 57 so that it points directly at the head while keeping the mic body and connecting cable out of the way. They also make the whole mic too, called a G5790.

For the bottom snare mic I'll use something that's very hyper-directional like a Sennheiser MD441U so that you're minimizing the leakage coming in from the kick drum and the bottom end of the kit.

On the kick drum lately I've been using the new DPA 4055 kick drum mic, but I'll also use a Shure Beta 52, or an AKG D12, D112. A Neumann U47 FET out front of the kick gives you that front head sound. It's pretty much the same thing that you always see, mainly because that's what top studios have.

I also used to lay a mic inside the kick through the hole in the front head and then carefully aim it at the beater for better isolation, or lay a Crown or AKG PZM microphone on the floor in front of the bass drum.

I like to use condensers on toms. Engineer Bill Schnee used to have four, Telefunken ELA M251s that he had modified with 20dB capsule pads just for toms. Boy, did they sound good. I was always so worried that the drummer was going to hit one of them!

You'll never have that in most situations so the next best thing would be Neumann U-87s. They have sort of a mid-range "bark" to them that's good on toms.

How about overheads?
Condensers again. I used to use Neumann KM84s or DPA 4011s but then I got into using these Milab DC96C condensers, which use a rectangular diaphragm so it sounds a little different and they pickup differently depending on orientation to the source. The diaphragm has a low mass so it's very articulated but not overly bright.

Do you use your overheads as cymbal mics or to record the whole kit?
They have to pick up the whole kit. In fact that's where I start with the drum sound. I move those around while the drummer plays because some drummers will nail a crash cymbal, which will make me move the mic away or tilt it so that cymbal isn't so loud.

The height of the overheads has everything to do with how live the room is. If you're working in a very live room then the overheads might have to be a little bit closer down than you would normally put them. Normally I'd put them about 3 feet over the toms.

I try to get the sound starting from the overheads and then add in the tom-toms and snare. I also monitor in mono pretty much the whole time to make sure there's not a polarity issue.

Do you have favorite preamps for drums?
You can't go wrong with good old Class-A Neve modules such as 1066, 1073, or 1272 but you want to make sure that you get ones that the capacitors are not all dried out so they're still able to pass low frequencies. I like API pre-amps and the S1P Sunset Sound "Tutti" are excellent. If you want tubes, there is the vintage Universal 610 and then Retro has a module with 3 tubes called 500Pre in a one slot 500-module that's great for drums.

Do you have a favorite vocal chain?
It depends on the singer. The microphone makes all the difference in the world so you have to start there. I've had great luck with a U47 if it's in great condition and quiet. An ELA M251 is brighter than a U47 but it's very nice too. For more modern stuff there are Neumann's new reissue U67 or M49V (both tube). There are a lot of great mics and not all are super-expensive either. Ask to try it out on your singer before purchase. It is an important decision for an artist/engineer/producer.

People always ask me the difference between the vintage and the reissue versions and what I've found is that the new ones are much quieter than the old vintage ones. Those original mics are pretty old now and can be unreliable and pretty noisy depending upon the state of the tube and the components inside. Why is that important? If you have a singer that sings softly, then the level of the noise floor becomes important and comes up if you have them sing up close to the mic and use a lot of pre-amp gain. On top of that, if you put a compressor and/or EQ after the preamp then all that will bring up more noise.

I take it that you use EQ or compression while you're recording.
Yes. You get a completely different result, and I believe a better result, when you're at least using some compression. If it's a pop recording, you're going to have to keep that vocal up front. I just find that everything works better when people are playing to or singing to a vocal that's a little bit steadier in level within the mix.

That said, I don't use very much compression. The needle barely moves and I set it up so there's only about 3 or 4dB of gain reduction at the loudest moment and using a low compression ratio.

I'm a little hesitant about committing to any kind of drastic EQ though. Sometimes you might want to go to a brighter mic instead, or add some midrange cut to a darker mic as long as it doesn't get too sibilant.

How about room mics?
I like to use omnidirectional condensers if it's a good sounding room. Sometimes ribbons are nice because they're a little duller and smoother and that can work out well if there are splashy hard surfaces in the room.

How do you record electric guitar?
I like to use two mics on the speaker, but where you place them on the speaker cone is super important. If you use a ribbon you may want to put it directly in the center over the dust cover. About halfway between the center of the speaker's dust cover and its edge surround is perfect for a dynamic mic like an SM7 or a 57.

A condenser mic is a little edgier and brighter so you might not have to EQ later. If you have the mic positioned correctly, then you won't have to do too much EQ later anyway. I find there is a sweet spot as you move the mic across the front of the speaker from near the surround suspension towards the dust cover. Sometimes, I would start with a mid-way position and just turn the mic slightly towards the dust cover to ever so slightly brighten the sound.

Do you have a favorite DI for bass?
I do like the passive Jensen DI with the Jensen transformer in it. I like an electronic active DI for a bass with passive pickups, like a low output P-Bass, because it can add a little bit of gain. If you already have an active bass then I wouldn't use an active DI though. A passive transformer box works better.

How about bass amp?
I like a condenser like a U87 or an EV RE20 dynamic but the choice of mic isn't critical for that.

There are a lot of approaches to miking acoustic guitar. Do you use one mic or two?
For the most part just one, but I'll move it around to find just the right spot. It depends on the part that's being played though. Finger picking versus strumming usually means two different locations. Strumming is much louder and can be very boomy if the mic is kept in a good spot that works for fingerpicking. I sometimes use a compressor to hold the level steady, especially for rhythm parts.

An X/Y pair of (Audio-Technica) AT-5054 cardioids or AT-4081 active ribbons about 10 inches from the sound hole can sound really good. It gives you some nice width, especially if the guitar is going to be in the center of your mix. If you double track it, you can use either a different pair of microphones or another stereo technique like mid-side.

I have to say that the acoustic guitar sound is really on the player. You get the best sounds with the best players and the best instruments. It's not that hard to record.

How do you handle leakage?
It's entirely dependent on where you're working. Sometimes it's perceived as a problem but that doesn't necessarily make it a problem. I think the two key areas to watch are vocalists in the same room as the band, or drums leaking into a vocal track.

If you don't have a lot of isolation booths then you're going to have to have all the players close together so that at least any leakage is pretty instant. In other words, instead of 20 or 30 milliseconds delay, it's much less.

Leakage is not necessarily a bad thing though. A lot of people like the leakage. I do if it's a good sounding room, but it all depends on the producer and what they might want to record over again.

What's the most difficult thing for you to record?
I would say some orchestral instruments. French horns can be problematic because of the nature of the sound. Certain string sections too, although that's not difficult, it's just a lot of work to get them to sound right.

How would you record a string section?
Back in the day we used to do 8-2-2 that's eight violins, two violas and two cellos, and double-track it. A bigger section might be 12-4-4, which is a nice sound. If you have the right size room that can support that many players then it sounds great. Of course a good arrangement is important too.

In the case of 8-2-2, we set up four violins in the first row with a mic between and over each pair of players. The second row would be the same. They all use the same model mic, which would be condensers like C414s or C12s but I've used U-87s, U47 FETs, or C12As. Similarly, I'd put a mic above and between the viola players and then a mic on each cello. A U87 has kind of a barky midrange that works well on cellos. I mike it over the bridge and the F-hole from about 3 feet away. Again if you have time, it's great to listen to each mic to make sure you're getting both musicians equally.

Again, mic distance is all about the room you're in and how live it is.

How about acoustic piano?

For live tracking sessions, I usually have the grand piano close to the drum kit so that any leakage would have no delay. I would set the lid on the low peg first and then I put the mics right over the hammers mainly to get it loud enough above the noise. You can move the mic over the treble strings down by the harp instead and that's a nice sound. For rock though, I'd put the mics right over the hammers and capture all that attack and bright sound so I don't have to EQ it too much later. The mic pre-amps make a noticeable difference on grand piano. I like Millennia Media's HV-3C, HV-37 or GML 8302 mic-pres.

Have you found any of the new clones of vintage mics that work for you?

I think there's a lot of hoopty-do about all this, but if the mic works for what you want it to do, then who cares what make or model it is? People will say you have to have a certain mic, but I think you have to look at these things one at a time, case by case. There's a little bit of a mystique about old microphones, but in the end it doesn't matter as long as the mic you have works for your application.

The fact of the matter is that I recorded a top ten record with $300 worth of drum mics. When we were doing Hall & Oats "Sarah Smile" in 1975, the producer and I were looking for a sound more like some of the old R&B records we always loved. So we recorded the drums all with SM57s. No one noticed that we didn't use the expensive mics, and no one cared. They just listened to the music, as it should be.

23

AL SCHMITT

After 20 Grammy's for Best Engineering (more than any other engineer) and work on over 150 gold and platinum records, Al Schmitt needs no introduction to anyone even remotely familiar with the recording industry. Indeed, his credit list is way too long to print here (but Henry Mancini, Paul McCartney, Neil Young, Bob Dylan, Steely Dan, George Benson, Toto, Natalie Cole, Quincy Jones, and Diana Krall are just some of them), but suffice it to say that Al's name is synonymous with the highest art that recording has to offer. Universally revered as one of the godfathers of modern recording, Al left us in 2021 but his recording wisdom thankfully lives on.

Do you use the same setup every time?

I usually start out with the same microphones. For instance, I know that I'm going to immediately start with a tube U 47 about 18 inches from the F-hole on an upright bass. That's basic for me and I've been doing that for years. I might move it up a little so it picks up a little of the finger noise. Now if I have a problem with a guy's instrument where it doesn't respond well to that mic then I'll change it, but that happens so seldom. Every once in a while I'll take another microphone and place it up higher on the fingerboard to pick up a little more of the fingering.

The same with the drums. There are times where I might change a snare mic or kick mic, but normally I use a D-112 or a 47 FET on the kick and a 451 or 452 on the snare and they seem to work for me. I'll use a Shure SM57 on the snare underneath and I'll put that microphone out of phase. I also mic the toms with 414's, usually with the pad in, and the hat with a Schoeps or a B&K or even a 451.

What are you using for overhead mics?

I do vary that. It depends on the drummer and the sound of the cymbals, but I've been using M 149's, the Royer 121's, or 451's. I put them a little higher than the drummer's head.

Do you try to capture the whole kit or just the cymbals?

I try to set it up so I'm capturing a lot of the kit, which makes it a little bigger sounding overall because you're getting some ambience.

What determines your mike selection?
It's usually the sound of the kit. I'll start out with the mics that I normally use and just go from there. If it's a jazz date then I might use the Royers and if it's more of a rock date then I'll use something else.

How much experimentation do you do?
Very little now. Usually I have a drum sound in 15 minutes so I don't have to do a lot. When you're working with the best guys in the world, their drums are usually tuned exactly the way they want and they sound great, so all you have to do is capture that sound. It's really pretty easy. And I work at the best studios where they have the best consoles and great microphones, so that helps.

I don't use any EQ when I record. I use the mics for EQ. I don't even use any compression. The only time I might use a little bit of compression is maybe on the kick, but for most jazz dates I don't.

How about mic preamps? Do you know what you're going to use? Do you experiment at all?
I know pretty much what I'm going to use. I have a rack of Neves that I'll use on the drums.

How do you handle leakage? Do you worry about it?
No, I don't. Actually leakage is one of your best friends because that's what makes things sometimes sound so much bigger. The only time leakage is a problem is if you're using a lot of crap mics. If you get a lot of leakage into them, it's going to sound like crap leakage, but if you're using some really good microphones and you're get some leakage, it's usually good because it makes things sound bigger.

I try to set everybody, especially in the rhythm section, as close together as possible. I come from the school when I first started where there were no headphones. Everybody had to hear one another in the room, so I still set up everybody up that way. Even though I'll isolate the drums, everybody will be so close that they can almost touch one another.

Let's talk about when you do an orchestra. Are you a minimalist, mic-wise?
Yes, I try to use a few as possible. On some of the dates I'll just use the room mics up over the conductors head. I'll have a couple of M 150's, or M 50's or even M 149's set to omnidirectional. I'll have some spot mics out there, but lots of times I don't even use those. It works if you have a conductor that knows how to bring out a section when it needs to be louder, so I'll just try to capture what he's hearing out there.

For violins I prefer the old Neumann U 67's. If I'm working on just violin overdubs I'll use the 67s and keep them in the omni position. I like the way that mic sounds when it's open and not in cardioid. It's much warmer and more open this way, but it's not always possible to do that because if there's brass playing at the same time then I'll just have to keep them in the cardioid position on the violins.

On violas, I like the Royer ribbon mics, the Neumann M 149s or the 67s, depending on availability. On celli I usually use the Neumann KM 84s or M 149s if they're available. The mics on the violins are about eight or ten feet above them; the same is true for the violas. For the celli, the mics may be 3 or 4 feet above them.

On harp, I like the Schoeps, the Royer or the Audio Technica 4060. On the French horns, I use the old M 49's. I use the M 149's on the rest of the woodwinds.

Do you have a philosophy in your approach when you're recording?

I get with the arranger, find out exactly what he's trying to accomplish, make sure that the artist is happy, and get the best sound I can possibly get on everything. If there's something that's near and dear to the artist or arranger, then I'll work towards pleasing them, although most of the time they're happy with what I get. Most of the guys that I work with, like Tommy LiPuma or David Foster, concentrate on the actual music and leave the sound up to me nine times out of ten.

I'm always very early on dates. I want to make absolutely sure that everything is working. I don't just click through mics, I talk into them to make sure that they sound right, then during the session I'm constantly out in the studio moving mics around until I get the sound that I'm happy with. I'll do this both between songs and every time there is a break.

What's the hardest thing for you to record?

Getting a great piano sound. The human voice is another thing that's tough to get. Other than that, things are pretty simple.

The larger the orchestra, the easier it is to record. The more difficult things are the eight and nine piece things, but I've been doing it for so long that none of it is difficult any more.

What mics do you use on piano?

I've been using the M 149s along with these old Studer valve preamps on piano, so I'm pretty happy with it lately. I try to keep them up as far away from the hammers as I can inside the piano. Usually one captures the low end and the other the high end and then I move them so it comes out as even as possible.

How about on vocals?

I try to keep the vocalist about six inches from the windscreen with the windscreen an inch or two from the mic, so the vocalist is anywhere from seven to ten inches from the microphone. That's usually a good place to start depending on the kind of sound you're looking for. If the vocalist is trying for a breathier quality, I'll move the mic up closer.

The microphone I'll use generally depends on the voice, the song, where it's being recorded, and the acknowledged favorite mic of the vocalist. For example, Barbra Streisand has been using this particular Neumann M 49 since we did "The Way We Were." It matches her voice so well that she will not use anything else. This particular mic is a rental, but she knows the specific serial number so that better be the right mic sitting up there when she's ready to record. That being said, I've done 12 song albums where I've used three different mics in the recording; one for up-tempo songs, one for medium tempo, and another on the ballads.

On Diana Krall and Natalie Cole I've used a special 67 treated by Klaus Heine into a Martech preamp, then I go into a Summit compressor where I pull about a dB or maybe two. I use very little compression, but I use it for the sound a lot. I also do a lot of hand compression as I record. I always have my hand on the vocal fader and ride the level to tape.

What's you're setup for horns?

I've been using a lot of 67's. On the trumpets I use a 67 with the pad in and I keep them in omnidirectional. I get them back about three or four feet off the brass. On saxophones I've been using M 149's. I put the mic somewhere around the bell so you can pick up some of the fingering. For clarinets, the mic should be somewhere up near the fingerboard and never near the bell.

For flute, I usually use a U67 positioned about three to four feet above the middle of the flute, but I may have to move it around a bit to find the sweet spot. If I want a tight sound, I may have the mic about 18 inches away. I may move it closer to the flautist's mouth or further down the fingerboard depending on the sound I'm trying to get. For flutes in a section, I usually have to get in a bit closer and more in front of the instrument.

How do determine the best place in the studio to place the instruments?

I'm working at Capital now and I've worked here so much that I know it like the back of my hand so I know exactly where to set things up to get the best sound. It's a given for me here. My setups stay pretty much the same. I try to keep the trumpets, trombones and the saxes as close as possible to one another so they feel like a big band, and I try to use as much of the room as possible.

I want to make certain the musicians are as comfortable as they can be with their setup. That means that they have clear sight lines to each other and are able to see, hear and talk to one another. This means having all the musicians as close together as possible. It facilitates better communication among them and that, in turn, fosters better playing.

To get a tight sound on the drums and to assure there's no leakage into the brass or string mics, I'll set the drums up in the drum booth. Then, I'll set the upright bass, the keyboard and the guitar near the drum booth so they all will be able to see and even talk easily to each other. Jazz setups generally involve small rhythm sections, so eye contact is critical. It's important that the bass player sees the piano player's left hand. Ideally they should all be close enough to almost be able to reach out and touch each other.

If there's a vocalist, 90 percent of the time I'll set them up in a booth. Very few choose to record in the open room with the orchestra, although Frank Sinatra and Natalie Cole come to mind.

On a large orchestral piece or a score for a motion picture, I set up the other instruments in the room as if I were setting up for a symphony orchestra. The violins are placed to the left, the violas in the center, and to the right will be the celli and the basses. Behind the violas will be the woodwinds and behind them the percussion, with French horns to left of center in the room and the other brass to the right of center.

If I am doing a big band setup, I'll put the saxophones to the left in the room and the trombones and trumpets to the right center. For a pop record, I will usually overdub these instruments.

If you had only one mic to use, what would it be?
A U 67. That's my favorite mic of all. I think it works well on anything. You can put it on a voice or an acoustic bass or an electric guitar, acoustic guitar, or a saxophone solo and it will work well. It's the jack of all trades and the one that works for me all the time.

GLOSSARY

5.1: The standard surround sound setup utilizing three speakers across the front and two in the rear.

0 VU: The nominal operating level when using equipment or plugins using a VU meter.

0dB FS: The highest level on a digital peak meter before overload. FS stands for Full Scale, which refers specifically to the digital meter.

ADC: Analog to Digital Converter. This device converts the analog waveform into the form of digital 1's and 0's.

AFL: After Fader Listen. A solo that listens to the audio that occurs after the channel fader in the signal path.

AIFF: Audio Interchange File Format (also known as "Apple Interchange File Format") is one of the most used audio file formats.

air. Frequencies above 10kHz that are more felt than heard. These frequencies can provide more realism to a sound used in the correct proportion.

ambience: The background noise of an environment.

API: Automated Process Incorporated; an American console manufacturer noted for their high-quality recording consoles.

arrangement: The way the instruments are combined in a song.

Atmos (see Dolby Atmos)

attack: The first part of a sound. On a compressor/limiter, a control that affects how that device will respond to the attack of a sound.

attenuation: A decrease in level.

automation: A system that memorizes, then plays back the position of all faders, mutes on a hardware console, and most other parameters in a DAW.

backline: The guitar, bass and keyboard amplifiers that stand at the back of the stage behind the musicians.

basic track: Recording the rhythm section for a record, which could be only the drums but could also include all the instruments of the band, depending upon the project.

bandwidth: The number of frequencies that a device will pass before the signal degrades. A human being can supposedly hear from 20Hz to 20kHz so the bandwidth of the human ear is 20Hz to 20kHz.

bi-directional: A microphone with a figure 8 pickup pattern.

binaural: A stereo recording technique using a model of a human head with microphones placed where the ears would be. This type of recording provides exceptional reproduction using headphones, but does not translate well to speakers.

Blumlein: A stereo miking configuration utilizing two figure 8 microphones.

bottom: Bass frequencies, the lower end of the audio spectrum. See also "low end"

bottom-end: See bottom.

bpm: Beats per minute. The measure of tempo.

buss: A signal pathway.

capacitor: An electronic component used to store energy. Because of its frequency bandwidth, it's a primary building block of analog filters and equalizers.

capsule: The part of a microphone that contains the primary electronic pickup element.

cardioid: A microphone that has a heart-shaped pickup pattern.

chorus: A type of signal processor where a detuned copy is mixed with the original signal, which creates a fatter sound.

clean: A signal with no distortion.

click: A metronome fed into the headphone mix to help the musicians play at the correct tempo.

clip: To overload and cause distortion.

clipping: When an audio signal begins to distort because a circuit in the signal path is overloaded, the top of the waveform becomes "clipped" off and begins to look square instead of rounded. This usually results in some type of distortion, which can be either soft and barely noticeable, or horribly crunchy sounding.

close miking: Placing a mic close to an instrument in order to decrease the pickup of room reflections or other sound sources.

coincident pair: A pair of the same model microphones placed with their capsules as close together as possible.

color: To affect the timbral qualities of a sound.

compression: Signal processing that controls and evens out the dynamics of a sound

compressor: A signal processing device used to compress audio dynamics.

condenser: A microphone that uses two electrically charged plates (thereby creating an electronic component known as a "condenser") as its basis of operation.

cue mix: The headphone mix sent to the musicians that differs from the one that the producer and engineer are listening to.

cut: To decrease, attenuate or make less.

DAC: Digital to Analog Converter. This device converts the digital 1's and 0's back to an analog waveform.

DAW: Digital Audio Workstation. A computer loaded with the appropriate hardware and software needed to record and edit audio.

dB: Stands for decibel, which is a unit of measurement of sound level or loudness. 1 dB is the smallest change in level that an average human can hear, according to many textbooks.

dBu: A measure of voltage in an analog audio system.

decay: The time it takes for a signal to fall below audibility.

Decca Tree: A microphone arrangement used primarily for orchestral recording that uses a spaced pair with a center mic connected to a custom stand and suspended over the conductor.

delay: A type of signal processor that produces distinct repeats (echoes) of a signal.

DI – Direct Inject; an impedance matching device (sometimes called a "direct box") for electric instruments that bypasses the use of a microphone.

diaphragm: The element of a microphone moved by sound pressure.

direct: To "go direct" means to bypass a microphone and connect the guitar, bass, keyboard, etc. directly into a recording device, usually through a direct box.

direct box: see DI.

directional: A microphone that has most of its pickup pattern in one direction.

Dolby Atmos. A surround sound technology developed by Dolby Laboratories. It expands on existing surround sound systems by adding height channels, allowing sounds to be interpreted as three-dimensional objects.

double: To play or sing a track a second time. The inconsistencies between both tracks make the part sound bigger.

dynamic: A dynamic microphone changes acoustic energy into electrical energy by the motion of a diaphragm through a magnetic field.

dynamic range: A ratio that describes the difference between the loudest and the quietest audio. The higher the number, equaling the greater dynamic range, the better

echo: For older engineers this is another word for reverb. For newer engineers this is another word for delay.

edgy: A sound with an abundance of mid-range frequencies.

electromagnetic induction: An electrical current that's produced by moving a conductor through a stationary magnetic field.

embouchure: The position and use of the lips, tongue, and teeth in playing a wind instrument, or the mouthpiece of a musical instrument.

EQ: Equalizer, or to adjust an equalizer (tone control) to affect the timbral balance of a sound.

equalizer: A tone control that can vary in sophistication from very simple to very complex (see parametric equalizer).

equalization: Adjustment of the frequency spectrum to even out or alter tonal imbalances.

feedback: When part of the output signal is fed back into the input.

FET: Field Effect Transistor; a solid state electronic component that has many of the same electronic qualities of a vacuum tube. Meant as a replacement for the vacuum tube, the FET has a much longer useful lifetime but lacks the sonic qualities.

figure 8: A microphone with a pickup pattern primarily from the front and rear, with very little on the sides.

flam: A sound source played slightly off-time with another.

flip the phase: Selecting the phase switch on a console, preamp or DAW channel in order to find the setting with the greatest bass response.

FS: Full scale. A digital peak meter that reads at 0 dB shows the full scale of the meter. The maximum amplitude of a digital system.

gain: The amount that a sound is boosted.

gain reduction: The amount of compression or limiting

gain staging: Setting the gain of each stage in the signal path so that one stage doesn't overload the next one in line.

gobo: A portable wall used to isolate one sound source from another.

groove: The pulse of the song and how the instruments dynamically breathe with it.

ground: A switch on some audio devices (mostly guitar amps and direct boxes) used to decrease hum.

headroom: The amount of dynamic range between the normal operating level and the maximum level, which is usually the onset of clipping.

Hz: An abbreviation for Hertz, which is the measurement unit of audio frequency, meaning the number of cycles per second. High numbers represent high sounds, and low numbers represent low sounds.

high end: The high frequency response of a device

high-pass filter: An electronic device that allows the high frequencies to pass while attenuating the low frequencies. Used to eliminate low frequency artifacts like hum and rumble. The frequency point where it cuts off can be fixed, switchable or variable.

impedance: The electronic measurement of the total electronic resistance to an audio signal.

I/O: The Input/Output of a device.

input pad: An electronic circuit that attenuates the signal, usually 10 or 20dB. See also "attenuation pad."

in the box: Doing all of your mixing with the software console in the DAW application on the computer, instead of using a hardware console.

iso booth: Isolation booth. An isolated section of the studio designed to eliminate leakage from coming in to the booth or leaking out.

intonation: The accuracy of tuning anywhere along the neck of a stringed instrument like a guitar or bass. Also applies to brass, woodwinds and piano.

key: An input on a noise gate that allows the gate to open up with the presence of a signal from another device or processor.

knee: The speed at which a compressor will turn on once it reaches threshold. A "soft knee" turns on gradually and is less audible than a "hard knee."

kHz: 1000 Hertz (example: 4kHz = 4,000Hz).

latency: Latency is a measure of the time it takes (in milliseconds) for the audio signal to pass through the computer during the recording process. This delay is caused by the time it takes for your computer to receive, understand, process, and send the signal back to your outputs.

lavaliere: A small microphone (sometimes called a "tie tac" or "lav") favored by broadcasters because of their unobtrusiveness.

layer: To make a larger more complex sound picture by adding additional tracks via overdubbing.

leakage: Sound from a distant instrument "bleeding" into a mic pointed at another instrument. Acoustic spill from a sound source other than the one intended for pickup.

Leslie: A speaker cabinet, usually used with a Hammond organ, that features rotating high and low frequency speakers.

LFE: Low Frequency Effects channel. The ".1" channel in a 5.1 surround system that has a bandwidth of about 30Hz to 120Hz.

limiter: A signal processing device used to constrict or reduce audio dynamics, reducing the loudest peaks in volume.

line level: The normal operating signal level of most professional audio gear. The output of a microphone is boosted to line level by a preamplifier.

look-ahead: In a digital limiter, look-ahead delays the audio signal a small amount (about two milliseconds) so that the limiter can anticipate the transients in such a way that it catches the peak before it gets by.

low-pass filter (LPF): A electronic frequency filter that allows only the low frequencies to pass while attenuating the high frequencies. The frequency point where it cuts off is usually either switchable or variable.

low end: The lower end of the audio spectrum, or bass frequencies usually below 200Hz.

make-up gain: A control on a compressor/limiter that applies additional gain to the signal. This is required since the signal is automatically decreased when the compressor is working. Make-up Gain "makes up" the gain and brings it back to where it was prior to being compressed.

Marshall cabinet: The most widely used guitar speaker cabinet. It contains four 12 inch speakers and is manufactured by Jim Marshall Amplifiers.

midrange: Middle frequencies starting from around 250Hz up to 4,000Hz.

mix buss: The network that mixes all of the individual channels together for your final mix.

modeling: Developing a software algorithm that is an electronic representation of the sound of hardware audio device down to the smallest behaviors and nuances.

modulate: The process of adding a control voltage to a signal source in order to change its character. For example, modulating a short slap delay with a 0.5Hz signal will produce chorusing.

mono: Short for monaural, or single audio playback channel.

monaural: A mix that contains a single channel and usually comes from only a one speaker.

M-S (Mid/Side): a stereo microphone technique utilizing directional and a figure 8 microphone.

muddy: Non-distinct because of excessive low frequencies.

mute: An on/off switch. To mute something would mean to turn it off.

nearfield: The listening area where there is more direct than reflected sound.

Neve: An English console manufacturer noted for its sonic qualities.

non-coincident pair: A stereo miking technique where two microphones are placed apart from one another at the distance approximately of your ears.

null: The point on the microphone pickup pattern where the pickup sensitivity is at its lowest.

Nyquist Theory: A basic theory of digital audio that states that the bandwidth of the audio is equal to half its sampling rate.

off-axis: A sound source away from the primary pickup point of a microphone.

omnidirectional: A microphone that picks up sound equally from any direction.

on-axis: A sound source aimed at the primary pickup point of a microphone.

ORTF: Office de Radiodif- fusion Television Française; a stereo miking technique developed by the Office of French Radio and Television Broadcasting using two cardioid mics angled 110 degrees apart and spaced seven inches (17 cm) apart horizontally.

overalls: In orchestral recording, the microphones that capture most of the sound.

overheads: The microphones placed over the head of a drummer used to either pick up the entire kit, or just the cymbals.

overs: Digital overs occur when the level is so high that it attempts to go beyond 0dB Full Scale on a typical digital level meter found in just about all digital equipment. A red Overload indicator usually will light, accompanied by a the crunchy, distorted sound of waveform clipping.

overdub: To record a new track while listening to previously recorded tracks.

overtone: The part of a sound that gives it its character and uniqueness.

outboard gear: Hardware devices like compressors, reverbs and effects boxes that are not built into a console and usually reside in an equipment rack in the control room.

out of phase: The polarity of two channels (it could be the left and right channel of a stereo program) are reversed, thereby causing the center of the program (like the vocal) to diminish in level. Electronically, when one cable is wired backwards from all the others.

pad: An electronic circuit that attenuates the signal (usually either 10 or 20dB) in order to avoid overload.

parametric equalizer: A tone control where the gain, frequency and bandwidth are all variable.

peaks: A sound that's temporarily much higher than the sound surrounding it.

PFL (Pre-Fader Listen): A solo that picks off the signal in the signal chain before it reaches the channel fader.

phantom image: In a stereo system, if the signal is of equal strength in the left and right channels, the resultant sound appears to come from in between them.

phase: The relationship between two separate sound signals when combined into one.

phase meter: A dedicated meter that displays the relative phase of a stereo signal.

phase shift: The process during which some frequencies (usually those below 100Hz) are slowed down ever so slightly as they pass through a device. This is usually exaggerated by excessive use of equalization and is highly undesirable.

plosive: A large puff of air that comes when a vocalist sings a word that a "p" sound.

plugin: An add-on to a computer application that adds functionality to it. EQ, modulation and reverb are examples of DAW plugins.

pop filter: A piece of acoustic foam, placed either internally near the diaphragm or externally over the mic, designed to reduce plosives, or "pops."

preamplifier: An electronic circuit that boosts the tiny output of a microphone to a level more easily used by the other electronic devices in the studio.

presence: Accentuated upper mid-range frequencies in the 5k to 10kHz range.

producer: The equivalent of a movie director, the producer has the ability to craft the songs of an artist or band technically, sonically and musically.

proximity effect: The inherent low frequency boost that occurs with a directional microphone as the signal source gets closer to it.

Pultec: An equalizer sold during the '50s and '60s by Western Electric that is highly prized today for its smooth sound.

pumping: When the level of a mix increases, then decreases noticeably. Pumping is caused by the improper setting of the attack and release times on a compressor.

punchy: A description for a quality of sound that infers good reproduction of dynamics with a strong impact. Sometimes means emphasis in the 200Hz and 5kHz areas.

PZM: Pressure Zone Microphone.

Q: The bandwidth of a filter or equalizer.

range: On a gate or expander, a control that adjusts the amount of attenuation that will occur to the signal when the gate is closed.

ratio: A parameter control on a compressor/limiter that determines how much compression or limiting will occur when the signal exceeds threshold.

release: The last part of a sound. On a compressor/limiter, a control that affects how that device will respond to the release of a sound.

resonance: See resonant frequency.

resonant frequency: A particular frequency or band of frequencies that are emphasized, usually due to some extraneous acoustic, electronic, or mechanical factor.

return: Inputs on a recording console especially dedicated for effects devices such as reverbs and delays. The return inputs are usually not as sophisticated as normal channel inputs on a console.

reverb: A type of signal processor that reproduces the spatial sound of an environment (i.e. the sound of a closet or locker room or inside an oil tanker).

RF: radio frequencies. Wireless audio systems are sometimes subject to dropouts caused by RF interference.

Rhodes: An electronic piano designed by Harold Rhodes and marketed by Fender in the 60s and 70s.

rhythm section: The instruments in a band that give the song it's pulse, usually the bass and drums.

ribbon: A microphone that utilizes a thin aluminum ribbon as the main pickup element.

roll off: Usually another word for high-pass filter, although it can refer to a low-pass filter as well.

rotor: The high frequency rotating speaker of a Leslie tone cabinet.

scope: Short for oscilloscope, an electronic measurement device that produces a picture of the audio waveform.

sibilance: A rise in the frequency response in a vocal where there's an excessive amount of 4k to 8kHz, resulting in the "S" sounds being overemphasized.

sidechain: A separate signal path to and from the control element of a dynamics device.

signal path: The electronic or digital circuitry that the audio signal must pass through.

solid state: Refers to electronic components, devices and systems based entirely on semiconductor materials such as silicon, germanium or gallium arsenide instead of vacuum tubes.

soundfield: The listening area containing mostly direct sound from the monitor speakers.

SoundField: A B-Format microphone for recording stereo or 5.1.

spaced pair: A stereo miking technique where the microphones are placed several feet apart.

source: An original master that is not a copy or a clone.

spectrum: The complete audible range of audio signals.

SPL: Sound Pressure Level. The air pressure caused by a sound wave.

spot mic: A microphone used during orchestral recording to boost the level of an instrument or soloist.

stage: In an analog console, a block of circuitry that performs a console function, like EQ or panning. In a digital or software console, a digital block that performs a console function.

subgroup: A separate sub-mixer that sums the assigned channels together, then sends that mix to the master mix buss.

sweetener: Another name for spot mic.

sympathetic vibration: vibrations, buzzes and rattles or notes that occur in areas of an instrument, or other instruments, other than the one that was struck.

talkback: The communication link between the control room and the cue mix in the musicians headphones allowing the producer or engineer to speak with the musicians.

threshold: The point at which an effect takes place. On a compressor/limiter, the Threshold control adjusts the point at which compression will begin to occur.

timbre: Tonal color.

trim: A control that sets the gain of a device, most usually on a microphone preamplifier.

track: A term sometimes used to mean a song. In recording, a separate musical performance that is recorded.

transformer: An electronic component that either matches or changes the impedance. Transformers are large, heavy and expensive but are in part responsible for the desirable sound in vintage audio gear.

transient: A very short duration signal.

tube: Short for vacuum tube; an electronic component used as the primary amplification device in most vintage audio gear. Equipment utilizing vacuum tubes run hot, are heavy, and have a short life, but have a desirable sound.

tunnel: A makeshift extension mounted to a bass drum used to isolate a microphone placed away from the drum head.

windscreen: A device placed over a microphone to attenuate the noise caused by wind interference.

X/Y: A stereo miking technique where the microphone capsules are mounted as closely as possible while crossing at 90 degrees.

ABOUT BOBBY OWSINSKI

Producer/engineer Bobby Owsinski is one of the best selling authors in the music industry with 25 books that are now staples in audio recording, music, and music business programs in colleges around the world, These include *The Mixing Engineer's Handboo*k, *The Music Mixing Workbook, Social Media Promotion For Musicians,* and more. He's also a contributor to Forbes writing on the new music business, his popular blogs and Inner Circle podcast have won numerous awards, and he's appeared on CNN and ABC News as a music branding and audio expert.

Visit Bobby's music production blog at bobbyowsinskiblog.com, his Music 3.0 music industry blog at music3point0.com, his podcast at bobbyoinnercircle.com, his online courses at bobbyowsinskicourses.com, and his website at bobbyowsinski.com.

BOBBY OWSINSKI BIBLIOGRAPHY

The Mixing Engineer's Handbook 5th Edition (BOMG Publishing)

The Music Mixing Workbook (BOMG Publishing)

The Recording Engineer's Handbook 5th Edition (BOMG Publishing)

The Mastering Engineer's Handbook 4th Edition (BOMG Publishing)

Social Media Promotion For Musicians 3rd Edition - *The Manual For Marketing Yourself, Your Band or your Music Online* (BOMG Publishing)

The Drum Recording Handbook 2nd Edition [with Dennis Moody] (Hal Leonard Publishing)

How To Make Your Band Sound Great (Hal Leonard Publishing)

The Studio Musician's Handbook [with Paul ILL] (Hal Leonard Publishing)

Music 4.1 - A Survival Guide To Making Music In The Internet Age 4th Edition (Hal Leonard Publishing)

The Music Producer's Handbook 2nd Edition (Hal Leonard Publishing)

The Musician's Video Handbook (Hal Leonard Publishing)

Mixing And Mastering With T-Racks: The Official Guide (Course Technology PTR)

The Touring Musician's Handbook (Hal Leonard Publishing)

The Ultimate Guitar Tone Handbook [with Rich Tozzoli] (Alfred Music Publishing)

The Studio Builder's Handbook [with Dennis Moody] (Alfred Music Publishing)

Abbey Road To Ziggy Stardust [with Ken Scott] (Alfred Music Publishing)

The Audio Mixing Bootcamp (Alfred Music Publishing)

Audio Recording Basic Training (Alfred Music Publishing)

Deconstructed Hits: Classic Rock Vol. 1 (Alfred Music Publishing)

Deconstructed Hits: Modern Pop & Hip-Hop (Alfred Music Publishing)

Deconstructed Hits: Modern Rock & Country (Alfred Music Publishing)

The PreSonus StudioLive Mixer Official Manual (Alfred Music Publishing)

You can get more info and read excerpts from each book by visiting the excerpts section of bobbyowsinski.com.

BOBBY OWSINSKI LINKEDIN LEARNING VIDEO COURSES

- The Audio Mixing Bootcamp Video Course
- Audio Recording Techniques
- Audio Mastering Techniques
- Music Studio Setup and Acoustics

BOBBY OWSINSKI ONLINE COURSES

Available at BobbyOwsinskiCourses.com

- Vocal Mixing Techniques
- The Music Mixing Primer
- Top 40 Mixing Secrets
- Music Mixing Accelerator
- Fully Booked

BOBBY OWSINSKI'S ONLINE CONNECTIONS

Website: bobbyowsinski.com

Courses: bobbyowsinskicourses.com

Podcast: bobbyoinnercircle.com

Music Production Blog: bobbyowsinskiblog.com

Music Industry Blog: music3point0.com

Forbes Blog: forbes.com/sites/bobbyowsinski/

Facebook: facebook.com/bobby.owsinski

YouTube: youtube.com/polymedia

Pinterest: pinterest.com/bobbyowsinski/

Linkedin: linkedin.com/in/bobbyo

Twitter: @bobbyowsinski

Time To Polish What You Recorded!
Add The Best Selling Book On Mixing Ever Written

Get 14% off when you order it at bobbyowsinski.com/handbook

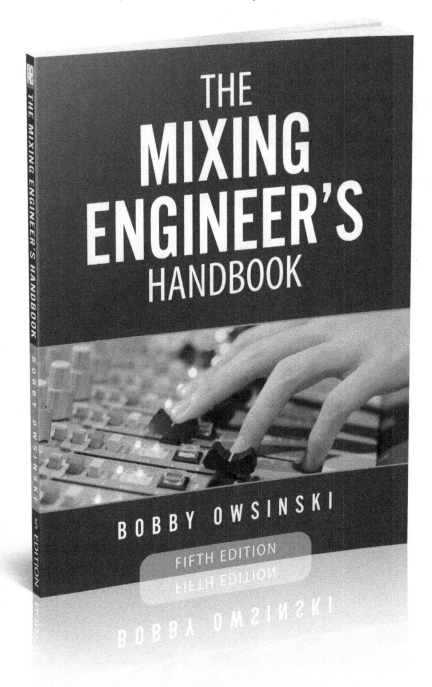

(Sorry, due to high shipping costs this offer is for customers in the United States only)

Thank so much for purchasing this book.
Here's an extra free bonus!

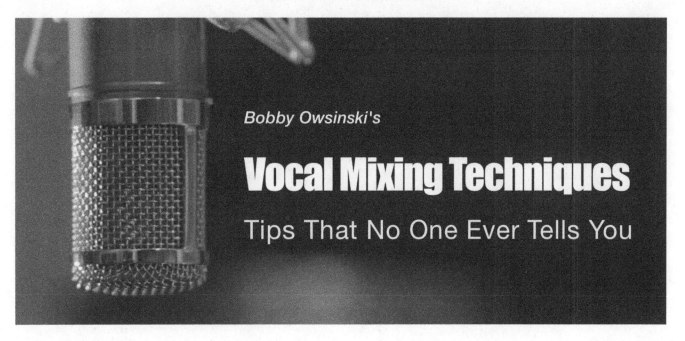

Includes 7 mixing technique videos for lead vocals, 2 for background vocals, plus a PDF summary, including:

- Vocal balance
- Vocal compression
- Vocal air
- Vocal effects
- Vocal pitch
- Background vocal panning
- Background vocal thickness

Plus these 2 bonus vocal recording techniques:

- How to set the best microphone distance
- How to get the best vocal performance

Download it here at bobbyowsinskicourses.com/vocal-bonus.

An Instructor Resource Kit for this book is available to qualified instructors

Everything you need to add *The Recording Engineer's Handbook* to your course right now!

Each kit includes:

- Syllabus
- Topics for Demonstrations and Discussions for each chapter
- Test Bank and answer key for 12 week semester
- Powerpoint and Keynote presentations for each chapter

"The Recording Engineer's Handbook Instructor Resource Kit is **free to qualified instructors using this book in their in music, recording or production courses.**

Send an email to office@bobbyowsinski.com to receive the download link.

www.ingramcontent.com/pod-product-compliance
Lightning Source LLC
Chambersburg PA
CBHW082213190225
22256CB00028B/719